Analytical and Computational Methods of Advanced Engineering Mathematics

Analytical and Computational Methods of Advanced Engineering Mathematics

Contributors

S. O. Edeki, H. I. Okagbue et al.

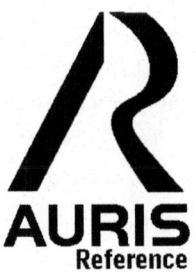

AURIS
Reference

www.aurisreference.com

Analytical and Computational Methods of Advanced Engineering Mathematics

Contributors: S. O. Edeki, H. I. Okagbue et al.

Published by Auris Reference Limited

www.aurisreference.com

United Kingdom

Copyright 2016
Printed in 2017 for Sale in the Indian Subcontinent

Analytical and Computational Methods of Advanced Engineering Mathematics

ISBN: 978-1-78154-820-2

British Library Cataloguing in Publication Data
A CIP record for this book is available from the British Library

Printed in the United Kingdom

Exclusively distributed by CBS Publishers & Distributors Pvt. Ltd.
Sales & Distribution Rights only for India, Pakistan, Bangladesh, Sri Lanka, Nepal and Bhutan.This book is not to be sold outside these territories.

Contents

List of Abbreviations

ADM	Adomian Decomposition Method
AGE	Alternating Group Explicit
AM	Arithmetic Mean
APD	Amplitude-Phase Distribution
BVP	Boundary Value Problem
DTM	Differential Transform Method
DVI	Differential Variational Inequalities
GE	Gaussian Elimination
GGE	Generalized Gaussian Elimination
HLI	Homogenous Linear Inequality
HLSF	Homogeneous Linear Feasibility Standard Form
HPM	Homotopy Perturbation Method
IVP	Initial Value Problem
LMM	Linear Multistep Method
MHD	Magnetohydrodynamic
MTSF	Mean Time to System Failure
NPC	NP-complete
ODE	Ordinary Differential Equation
SFEM	Stochastic Finite Element Method
TPBVP	Two-Point Boundary Value Problem
VS	Variable Substitution

List of Contributors

S. O. Edeki
Department of Mathematics, College of Science & Technology, Covenant University, Otta, Nigeria

H. I. Okagbue
Department of Mathematics, College of Science & Technology, Covenant University, Otta, Nigeria

A. A. Opanuga
Department of Mathematics, College of Science & Technology, Covenant University, Otta, Nigeria

S. A. Adeosun
Department of Mathematical Sciences, Crescent University, Abeokuta, Nigeria

Vembu Ananthaswamy
Department of Mathematics, The Madura College, Madurai, India

Lakshmanan Rajendran
Department of Mathematics, The Madura College, Madurai, India

Petro Savenko
Department of Numerical Methods of the Pidstryhach Institute for Applied Problems of Mechanics and Mathematics, National Academy of Sciences of Ukraine, Lviv, Ukraine

Olabode B. T
Department of Mathematical Sciences, Federal University of Technology, Akure, Nigeria

Momoh A. L.
Department of Mathematical Sciences, Federal University of Technology, Akure, Nigeria

Yusuf A
Department of Mathematics and Statistics, Federal University of Technology, Minna, Nigeria

Aiyesimi Y. M.
Department of Mathematics and Statistics, Federal University of Technology, Minna, Nigeria

Jiya M.
Department of Mathematics and Statistics, Federal University of Technology, Minna, Nigeria

Okedayo G. T
Department of Mathematics, Ondo State University of Science and Technology, Okiti-Pupa, Nigeria

Drakos Stefanos
International Centre for Computational Engineering, Rhodes, Greece

Paul T. R. Wang
Wang Paul_Research, Potomac, Maryland, USA

Emanuele Galligani
Department of Mathematics "G. Vitali", University of Modena and Reggio Emilia, Via Campi 213/b, I-41125, Modena, Italy

Ibrahim Yusuf
Department of Mathematical Sciences, Faculty of Science, Bayero University, Kano, Nigeria

Nafiu Hussaini
Department of Mathematical Sciences, Faculty of Science, Bayero University, Kano, Nigeria

Preface

The text *Analytical and Computational Methods of Advanced Engineering Mathematics* focuses on the topics which are an essential part of the engineering mathematics course: ordinary differential equations, vector calculus, linear algebra and partial differential equations. First chapter presents a semi-analytical method (DTM) for solving a certain class of ordinary differential equations (ODEs). In second chapter, we apply Homotopy perturbation method (HPM) to the nonlinear Bratu's problem, Troesch's problem, and catalytic reactions in flat particles. Computational methods in the theory of synthesis of radio and acoustic radiating systems have been discussed in third chapter. Fourth chapter proposes continuous hybrid multistep methods with Legendre polynomial as basic functions for the direct solution of system of second order ordinary differential equations. The analysis of couette flow of a nanofluid in an inclined channel with soret and dufour effects has been carried out in fifth chapter. Sixth chapter presents a procedure of conducting stochastic finite element analysis using polynomial chaos in order to propagate the uncertainties of input to constitutive relation of stress and strain. Seventh chapter generalizes the traditional Gaussian elimination (GE) technique to resolve the feasibility of any system of linear inequalities or equalities. Eighth chapter deals with the solution of nonlinear system arising from finite difference discretization of nonlinear diffusion convection equations by the lagged diffusivity functional iteration method combined with different inner iterative solvers. Ninth chapter discusses the stochastic modeling of repairable redundant system comprising one big unit and three small dissimilar units. Last chapter concerns with the analysis of the variation of the solution of some optimal control problems as the number of discretization points increases.

Chapter 1

A SEMI-ANALYTICAL METHOD FOR SOLUTIONS OF A CERTAIN CLASS OF SECOND ORDER ORDINARY DIFFERENTIAL EQUATIONS

S. O. Edeki[1], H. I. Okagbue[1], A. A. Opanuga[1], S. A. Adeosun[2]

[1]Department of Mathematics, College of Science & Technology, Covenant University, Otta, Nigeria

[2]Department of Mathematical Sciences, Crescent University, Abeokuta, Nigeria

ABSTRACT

This paper presents the theory and applications of a new computational technique referred to as Differential Transform Method (DTM) for solving second order linear ordinary differential equations, for both homogeneous and nonhomogeneous cases. For the robustness and efficiency of the method, four examples are considered. The results indicate that the DTM is reliable and accurate when compared to the exact solutions of the solved problems.

INTRODUCTION

Most of the problems encountered in applied sciences, management and economics take the forms of second order linear ordinary differential equations. Sometimes, obtaining exact solutions through the direct methods (analytical methods) for these systems seems difficult even if the exact solutions exist, hence the need for numerical techniques for approximate solutions. Some of these numerical methods involve linearization, disscretization and perturbation, and they only permit the solutions to a given ODE at a certain interval. In addition, they are intensive in terms of computation and as such, lead to the situations where some basic phenomena are technically avoided.

The notion of DTM was first introduced by Zhou [1] while solving linear and nonlinear initial value problems in electric circuit analysis. The Method provides an analytical approximate solution to linear and nonlinear systems of differential equations. This involves the construction of a semi-analytical technique with the aid of a Taylor series expansion for the solutions in polynomial forms. The DTM has been proven by many researchers [1] -[7] to be very effective and efficient in handling differential equations with point boundary value problems, kvd and mkdv differential-algebraic equations, and so on. It also reduces the size of computational work while maintaining high accuracy even with a fast convergence rate compared to the theoritical solutions [8] .

Some other analytical methods like the variational iterative method and the Adomian decomposition method witness difficulties in handling functions involving complicated integrals; since successive component using those methods depends on the previous component [9] -[13] but DTM overcomes this by solving an algebraic recursive equation.

THE DIFFERENTIAL TRANSFORM METHOD

This section introduces the basic concepts and theorems of DTM needed for applications in the remaining sections.

Definition 1. Let $f(x)$ be a given function of one variabe defined at a point $x = x_0$, then the one-dimensional k^{th} differential transform of the $f(x)$ defined as $F(k)$ is given as:

$$F(k) = \frac{1}{k!}\left(\frac{d^k f(x)}{dx^k}\right)\Bigg|_{x=x_0} \quad (1)$$

Equation (1) is the transformed function of $f(x)$.

Definition 2. The differential inverse transform of $F(k)$ is a Taylor series expansion of the function $f(x)$ about $x = x_0 = 0$, defined as :

$$f(x) = \sum_{k=0}^{\infty} F(k)x^k \quad (2)$$

Combining (1) and (2) yields:

$$f(x) = \sum_{k=0}^{\infty}\left(\frac{d^k f(x)}{dx^k}\right)\frac{x^k}{k!} \quad (3)$$

Some Basic Theorems of the Differential Transform Method

The following theorems and properties of the DTM are stated below for the sake of applications, their proofs can be found in [14] and [15].

Let $f_1(x)$, $f_2(x)$ and $f_*(x)$ be differentiable functions with differential transforms $F_1(k)$, $F_2(k)$ and $F_*(k)$ respectively, with, $\alpha \in \mathbb{R}$ and δ a Kronecker delta, then the following theorems hold:

Theorem 2.1 If $y = f_1(x) \pm f_2(x)$ then $Y(k) = F_1(k) \pm F_2(k)$

Theorem 2.2 If $y = \alpha f_*(x)$ then $Y(k) = \alpha F_*(k)$

Theorem 2.3 If $y = x^n$ then $Y(k) = \delta(k-n)$ such that:

$$Y(k) = \delta(k-n) = \begin{cases} 1, & k = n \\ 0, & \text{otherwise} \end{cases}$$

Theorem 2.4 If $y = e^{\alpha x}$ then $Y(k) = \dfrac{\alpha^k}{k!}$, where α is a constant.

Theorem 2.5 If $y = \dfrac{d^n}{dx^n}[f_*(x)]$ then $Y(k) = \dfrac{(k+n)!}{n!} F_*(k+n)$.

In particular, we have:

(a) If $y = \dfrac{d}{dx}[f_*(x)]$ then $Y(k) = (k+1)F_*(k+1)$

(b) If $y = \dfrac{d^2}{dx^2}[f_*(x)]$ then $Y(k) = (k+1)(k+2)F_*(k+2)$

The DTM and the Second order Linear Ordinary Differential Equations (ODEs)

In this section, we present clearly how a second order linear ODE with constant coefficients is transformed using the DTM.

The corresponding ODE is of the form:

$$ay''(x) + by'(x) + cy(x) = f(x), \quad a \neq 0 \tag{4}$$

with initial conditions $y(0) = \alpha_1$ and $y'(0) = \alpha_2$

Equation (4) is re-expressed in the form of

$$y''(x) = g(x, y(x), y'(x)) \tag{5}$$

Thus, (4) becomes

$$y''(x) = \frac{1}{a}\left[f(x) - cy(x) - by'(x)\right]$$

(6)

We will take the differential transform (DT) of (6) by applying theorems (2.1-2.6) as follows:

$$DT\left[y''(x) = \frac{1}{a}\left[f(x) - by'(x) - cy(x)\right]\right]$$

(7)

$$(k+1)(k+2)Y(k+2) = \frac{1}{a}\left[F(x) - b(k+1)Y(k+1) - cY(k)\right]$$

(8)

$$Y(k+2) = \frac{1}{a(k+1)(k+2)}\left[F(x) - b(k+1)Y(k+1) - cY(k)\right]$$

(9)

subject to the initial conditions $Y(0) = Y_0 = \alpha_1$ and $Y'(0) = Y(1) = \alpha_2$.

Equation (9) is a recursive formula for the computation of coefficient terms in the series solution of the problem. Therefore, using (2) and (9) gives the approximate solution of (4) as:

$$y(x) = \sum_{k=0}^{\infty} Y(k)x^k$$

(10)

APPLICATIONS AND NUMERICAL RESULTS

In this section, we will apply the discussed DTM to solve some problems whose results will be compared with the theoretical (exact) solutions. Two cases with two examples each are considered. Case I and Case II for homogeneous and nonhomogeneous respectively.

Case I Example 1: Consider the ODE

$$y''(x) - 4y'(x) + 3y(x) = 0$$

(11)

subject to $y(0) = 1$, $y'(0) = 1$ with a theoretical solution

$$y(x) = e^x$$

(12)

Procedure We rewrite (11) in a standard form and take the differential transform

(DT) as follows:

$$DT\left[y''(x)=4y'(x)-3y(x)\right]_{(13)}$$

$$\Rightarrow (k+1)(k+2)Y(k+2)=4(k+1)Y(k+1)-3Y(k) \tag{14}$$

$$\therefore Y(k+2)=\frac{1}{(k+1)(k+2)}\left[4(k+1)Y(k+1)-3Y(k)\right] \tag{15}$$

with the initial conditions $Y(0)=1,Y(1)=1$.

By using the recursive relation in (15) with $k\geq 0$, we obtain values for $Y(2),Y(3),Y(4),\cdots$, as showed below:

For $k=0,Y(2)=\frac{1}{2!}$; for $k=1,Y(3)=\frac{1}{3!}$; for $k=2,Y(4)=\frac{1}{4!}$; for $k=3,Y(5)=\frac{1}{5!}$, $\bullet\bullet\bullet$; for $k=n,Y(n)=\frac{1}{n!}$

But from (10),

$$y(x)=\sum_{k=0}^{\infty}Y(k)x^k$$

$$y(x)=Y(0)+Y(1)x+Y(2)x^2+Y(3)x^3+Y(4)x^4+\cdots$$

$$=1+x+\frac{x^2}{2!}+\frac{x^3}{3!}+\frac{x^4}{4!}+\frac{x^5}{5!}+\cdots=e^x \tag{16}$$

Equation (16) is the same with the theoretical solution in (12).

Case I Example 2: Consider the ODE

$$y''(x)+y(x)=0 \tag{17}$$

subject to $y(0)=1$ and $y'(0)=0$, with a theoretical solution

$$y(x)=\cos x \tag{18}$$

Procedure We re-write (16) in a standard form and take the differential transform as follows:

$$DT\left[y''(x)=-y(x)\right]$$

$$\therefore (k+1)(k+2)Y(k+2)=-Y(k)$$

$$\Rightarrow Y(k+2)=\frac{-Y(k)}{(k+1)(k+2)} \tag{19}$$

with the initial conditions $Y(0)=1, Y(1)=0$.

Computing $Y(2), Y(3), Y(4), \cdots$, for $k \geq 0$ using (19) gives the following:

For $k=0, Y(2)=\dfrac{-1}{2!}$; for $k=1, Y(3)=0$; for $k=2, Y(4)=\dfrac{1}{4!}$; for $k=3, Y(5)=0$,

for $k=4, Y(6)=\dfrac{-1}{6!}$, $\bullet\bullet\bullet$; for $k=n, Y(2n+2)=\dfrac{(-1)^n}{(2n+2)!}$

We observed that $Y(1)=Y(3)=Y(5)=\cdots=Y(2n+1)=0$ for $k \geq 0$. Hence, the solution of (17) by (10) is reformed as:

$$y(x)=\sum_{k=0}^{\infty} Y(2k)x^{2k} = Y(0)+Y(2)x^2 +Y(4)x^4 +Y(6)x^6 \cdots$$

$$=1-\frac{x^2}{2!}+\frac{x^4}{4!}-\frac{x^6}{6!}+\cdots = \cos x \tag{20}$$

Equation (20) agrees with the theoritical solution in (18).

Case II Example 1: Consider the ODE

$$y''(x)-y(x)=x \tag{21}$$

subject to $y(0)=1$ and $y'(0)=0$, with a theoretical solution:

$$y(x)=e^x - x \tag{22}$$

Procedure Equation (21) by the differential transform method becomes;

$$DT\left[y''(x)= y(x)+x\right]$$

$$\Rightarrow Y(k+2)=\frac{1}{(k+1)(k+2)}\left[Y(k)+\delta(k-1)\right] \tag{23}$$

with the initial conditions $Y(0)=1, Y(1)=0$.

Thus, for $k \geq 0$, the values of $Y(2), Y(3), Y(4), \cdots$, obtained using (23) are given below:

For $k=0, Y(2)=\dfrac{1}{2!}$; for $k=1, Y(3)=\dfrac{1}{3!}$; for $k=2, Y(4)=\dfrac{1}{4!}$; for $k=3, Y(5)=\dfrac{1}{5!}$, for

$\bullet\bullet\bullet$ $k=4, Y(6)=\dfrac{1}{6!}$; for $k=n, Y(n)=\dfrac{1}{n!}$.

As such, from (10), the solution of (21) is:

$$y(x) = \sum_{k=0}^{\infty} Y(k) x^k = Y(0) + Y(1)x + Y(2)x^2 + Y(3)x^3 + Y(4)x^4 + \cdots$$

$$= 1 + 0 + \frac{x^2}{2!} + \frac{x^3}{3!} + \frac{x^4}{4!} + \frac{x^5}{5!} + \cdots = 1 + \sum_{n=0}^{\infty} \frac{x^n}{n!}$$

(24)

Since

$$\sum_{n=0}^{\infty} \beta^n = \sum_{n=0}^{v} \beta^n + \sum_{n=v+1}^{\infty} \beta^n$$

(25)

$$\therefore \quad y(x) = 1 + \sum_{n=2}^{\infty} x^n = -x + e^x$$

(26)

Equation (26) is the same with the theoretical solution in (22).

Case II Example 2: Consider the ODE

$$y''(x) - 3y'(x) + 2y(x) = e^x$$ (27)

subject to $y(0) = 1$ and $y'(0) = 0$, with a theoretical solution:

$$y(x) = (1-x)e^x$$ (28)

Procedure Equation (27) by the differential transform method becomes;

$$DT\left[y''(x) - 3y'(x) + 2y(x) = e^x \right]$$

$$\Rightarrow Y(k+2) = \frac{1}{(k+1)(k+2)} \left[3(k+1)Y(k+1) - 2Y(k) + \frac{1}{k!} \right]$$ (29)

subject to the initials $Y(0) = 1, Y(1) = 0$.

Thus, for $k \geq 0$, the values of $Y(2), Y(3), Y(4), \cdots$, obtained using (29) are given below:

For $k = 0, Y(2) = \frac{-1}{2!}$; for $k = 1, Y(3) = \frac{-2}{3!}$; for $k = 2, Y(4) = \frac{-3}{4!}$; for $k = 3, Y(5) = \frac{-4}{5!}$,

for $k = 4, Y(6) = \frac{1}{6!}$; $\bullet\bullet\bullet$; for $k = n, Y(n) = \frac{-(n-1)}{n!} = \frac{1}{n!} - \frac{1}{(n-1)!}$.

Hence, from (10), the solution of (27) is:

$$y(x) = \sum_{k=0}^{\infty} Y(k)x^k = Y(0) + Y(1)x + Y(2)x^2 + Y(3)x^3 + Y(4)x^4 + \cdots$$

$$= 1 + 0 - \frac{x^2}{2!} - \frac{2x^3}{3!} - \frac{3x^4}{4!} - \frac{4x^5}{5!} - \frac{5x^5}{6!} \cdots = 1 - \sum_{n=2}^{\infty} (n-1)\frac{x^n}{n!}$$

$$= -\sum_{n=0}^{\infty} (n-1)\frac{x^n}{n!} = \sum_{n=0}^{\infty} \frac{x^n}{n!} - x\sum_{n=1}^{\infty} \frac{x^{n-1}}{(n-1)!}$$

$$= \sum_{n=0}^{\infty} \frac{x^n}{n!} - x\sum_{m=0}^{\infty} \frac{x^m}{(m)!}, \text{ for } m = n-1$$

$$\therefore \ y(x) = e^x - xe^x \tag{30}$$

Equation (30) corresponds with the theoretical solution in (28).

Remark 3.1:

We present numerical comparisons between the exact, and the numerical solutions based on a 5-iterate DTM using the coefficient terms from Table 1, Table 2, Table 3, and Table 4 as shown inTable 5, Table 6, Table 7, and Table 8 respectively.

DISCUSSION OF RESULTS AND CONCLUSION

In this paper, we have presented a semi-analytical method (DTM) for solving a certain class of ODEs. The DTM has advantages over other numerical techniques as it does not involve linearization, discretization or perturbation of a given problem; hence it has no effect of computational round off error. The DTM also provides a closed-form solution; therefore, it is very powerful and effective in finding both analytical and numerical solutions of second order linear ODEs with constant coefficients.

Table 1: Coefficient terms from Equation (15).

k	0	1	2	3	4	...	n
$Y(k+2)$	$Y(2) = \frac{1}{2!}$	$Y(3) = \frac{1}{3!}$	$Y(4) = \frac{1}{4!}$	$Y(5) = \frac{1}{5!}$	$Y(6) = \frac{1}{6!}$	...	$Y(n) = \frac{1}{n!}$

Table 2: Coefficient terms from Equation (19).

k	0	1	2	3	4	...	n
$Y(k+2)$	$Y(2) = \frac{-1}{2!}$	$Y(3) = 0$	$Y(4) = \frac{1}{4!}$	$Y(5) = 0$	$Y(6) = \frac{-1}{6!}$	...	$Y(2n+2) = \frac{(-1)^n}{(2n+2)!}$

Table 3. Coefficient terms from Equation (23).

k	0	1	2	3	4	...	n
$Y(k+2)$	$Y(2)=\dfrac{1}{2!}$	$Y(3)=\dfrac{1}{3!}$	$Y(4)=\dfrac{1}{4!}$	$Y(5)=\dfrac{1}{5!}$	$Y(6)=\dfrac{1}{6!}$	...	$Y(n)=\dfrac{1}{n!}$

Table 4: Coefficient terms from Equation (29).

k	0	1	2	3	4	...	n
$Y(k+2)$	$Y(2)=\dfrac{-1}{2!}$	$Y(3)=\dfrac{-2}{3!}$	$Y(4)=\dfrac{-3}{4!}$	$Y(5)=\dfrac{-4}{5!}$	$Y(6)=\dfrac{1}{6!}$	...	$Y(n)=\dfrac{-(n-1)}{n!}$

Table 5: Numerical comparisons for Case I Example 1.

x	Exact solution	5-Iterate DTM	Absolute Error
0.0	1.000000000	1.000000000	0.000000000
0.1	1.105170918	1.105170917	1.40898E−09
0.2	1.221402758	1.221402667	9.14935E−08
0.3	1.349858808	1.349857750	1.05758E−06
0.4	1.491824698	1.491818667	6.03097E−06
0.5	1.648721271	1.648697917	2.3354E−05
0.6	1.822118800	1.822048000	7.08004E−05
0.7	2.013752707	2.013571417	0.000181291
0.8	2.225540928	2.225130667	0.000410262
0.9	2.459603111	2.458758250	0.000844861
1.0	2.718281828	2.716666667	0.001615162

Table 6: Numerical comparisons for Case I Example 2.

x	Exact solution	5-Iterate DTM	Absolute Error
0.0	1.000000000	1.00000000	0.00000000
0.1	0.995004165	0.99500418	1.52775E−08
0.2	0.980066578	0.98006711	5.3327E−07
0.3	0.955336489	0.95534088	4.38587E−06
0.4	0.921060994	0.92108089	1.98949E−05
0.5	0.877582562	0.87764757	6.50076E−05
0.6	0.825335615	0.82550800	0.000172385
0.7	0.764842187	0.76523760	0.000395410
0.8	0.696706709	0.69752178	0.000815068
0.9	0.621609968	0.62315763	0.001547657
1.0	0.540302306	0.54305556	0.002753250

Table 7: Numerical comparisons for Case II Example 1.

x	Exact solution	5-Iterate DTM	Absolute Error
0.0	1.000000000	1.000000000	0.000000000
0.1	1.005170918	1.005170917	1.40898E-09
0.2	1.021402758	1.021402667	9.14935E-08
0.3	1.049858808	1.049857750	1.05758E-06
0.4	1.091824698	1.091818667	6.03097E-06
0.5	1.148721271	1.148697917	2.3354E-05
0.6	1.222118800	1.222048000	7.08004E-05
0.7	1.313752707	1.313571417	0.000181291
0.8	1.425540928	1.425130667	0.000410262
0.9	1.559603111	1.558758250	0.000844861
1.0	1.718281828	1.716666667	0.001615162

Table 8: Numerical comparisons for Case II Example 2.

x	Exact solution	5-Iterate DTM	Absolute Error
0.0	1.000000000	1.000000000	0.000000000
0.1	0.994653826	0.994653840	1.40097E-08
0.2	0.977122207	0.977123111	9.04583E-07
0.3	0.944901165	0.944911563	1.03972E-05
0.4	0.895094819	0.895153778	5.89592E-05
0.5	0.824360635	0.824587674	0.000227038
0.6	0.728847520	0.729532000	0.000684480
0.7	0.604125812	0.605868840	0.001743028
0.8	0.445108186	0.449031111	0.003922925
0.9	0.245960311	0.253995063	0.008034751
1.0	0.000000000	0.015277778	0.015277778

REFERENCES

1.   Zhou, J.K. (1986) Differential Transformation and its Applications for Electrical Circuits. Huazhong University Press, Wuhan.

2.   Chen, C.L. and Liu, Y.C. (1988) Solution of Two Point Boundary Value Problems Using the Differential Transformation Method. Journal of Optimization Theory and Applications, 99, 23-35. http://dx.doi.org/10.1023/A:1021791909142

3.   Ayaz, F. (2004) Application of Differential Transform Method to Differential-algebraic Equations. Applied Mathematics and Computation, 152, 649-657. http://dx.doi.org/10.1016/S0096-3003(03)00581-2

4.   Kangalgil, F. and Ayaz, F. (2009) Solitary Wave Solutions for the kdv and mkdv Equations by Differential Transform Method. Chaos, Solitons and Fractals, 41, 464-472. http://dx.doi.org/10.1016/j.chaos.2008.02.009

5. Batiha, K. and Batiha, B. (2011) A New Algorithm for Solving Linear Ordinary Differential Equations. World Applied Sciences Journal, 15, 1774-1779.

6. Thongmoon, M. and Pusjuso, S. (2010) The Numerical Solutions of Differential Transform Method and the Laplace Transform Methods for a System of Differential Equations. Nonlinear Analysis: Hybrid Systems, 4, 425-431. http://dx.doi.org/10.1016/j.nahs.2009.10.006www.elsevier.com/locate/nahs

7. Ravi Kanth, A.S.V. and Aruna, K. (2009) Differential Transform Method for Solving the Linear and Nonlinear Klein-Gordon Equation. Computer Physics Commmunications, 180, 708-711. http://dx.doi.org/10.1016/j.cpc.2008.11.012

8. Chang, S.H. and Chang, I.L. (2008) A New Algorithm for Calculating One-Dimensional Differential Transform of Nonlinear Functions. Applied Mathematics and Computation, 195, 799-808. http://dx.doi.org/10.1016/j.amc.2007.05.026

9. Adomian, G. (1994) Solving Frontier Problems of Physics. The Decomposition Method. Springer, New York.

10. Ibijola, E.A. and Adegboyegan, B.J. (2008) On the Theory and Application of Adomian Decomposition Method for Solution of Second Order ODEs. Pacific Journal of Science and Technology, 9, 357-362.

11. Shousa, D.-H. and He, J.-H. (2008) Beyond Adomian Method: The Variational Iteration Method for Solving Heat-Like and Wave-Like Equations with Variable Coefficients. Physics Letters A, 372, 233-237. http://dx.doi.org/10.1016/j.physleta.2007.07.011

12. He, J.-H. (2007) Variational Iterative Method—Some Recent Results and New Interpretations. Journal of Computational and Applied Mathematics, 207, 3-17. http://dx.doi.org/10.1016/j.cam.2006.07.009

13. Catal, S. (2012) Some of Semi Analitical Methods for Blasius Problem. Applied Mathematics, 3, 727-728. http://dx.doi.org/10.4236/am.2012.37106

14. Arikoglu, A. and Ozkol, I. (2005) Solution of Boundary Value Problems for Integro-Differential Equations by Using Transform Method. Applied Mathematics and Computation, 168, 1145-1158. http://dx.doi.org/10.1016/j.amc.2004.10.009

15. Odibat, Z. (2008) Differential Transform Methods for Solving Volterra Integral Equations with Separable Kernels. Mathematical and Computer Modelling, 48, 1144-1146. http://dx.doi.org/10.1016/j.mcm.2007.12.022

Chapter 2

ANALYTICAL SOLUTIONS OF SOME TWO-POINT NON-LINEAR ELLIPTIC BOUNDARY VALUE PROBLEMS

Vembu Ananthaswamy, Lakshmanan Rajendran

Department of Mathematics, The Madura College, Madurai, India

ABSTRACT

Several problems arising in science and engineering are modeled by differential equations that involve conditions that are specified at more than one point. The non-linear two-point boundary value problem (TPBVP) (Bratu's equation, Troesch's problems) occurs engineering and science, including the modeling of chemical reactions diffusion processes and heat transfer. An analytical expression pertaining to the concentration of substrate is obtained using Homotopy perturbation method for all values of parameters. These approximate analytical results were found to be in good agreement with the simulation results.

INTRODUCTION

All chemical reactions are usually accompanied with mass and energy transfer, either homogeneously or heterogeneously. Mathematical modeling for these processes is based on material and energy balance. One can generate a set of differential equations known as the reaction-diffusion problem. Owing to the strong nonlinearity of the reaction rate, mainly from the effect of temperature, reaction-diffusion equations are paid more attention in analyzing and designing chemical and catalytic reactors [1]. The same phenomena exist in electrochemical processes, with the add complexity of a varying potential field, and considerable research has been reviewed for electrochemical reactions occurring in the porous electrode [2].

Linear and nonlinear phenomena are of fundamental importance in various fields of science and engineering. Most models of real-life problems are still very difficult to solve. Therefore, approximate analytical solutions such as Homotopy perturbation method (HPM) [3-12] were introduced. This method is the most effective and convenient ones for both linear and nonlinear equations. Perturbation method is based on assuming a small parameter. The majority of nonlinear problems, especially those having strong nonlinearity, have no small parameters at all and the approximate solutions obtained by the perturbation methods, in most cases, are valid only for small values of the small parameter. Generally, the perturbation solutions are uniformly valid as long as a scientific system parameter is small. However, we cannot rely fully on the approximations, because there is no criterion on which the small parameter should exists. Thus, it is essential to check the validity of the approximations numerically and/or experimentally. To overcome these difficulties, HPM have been proposed recently. In this paper we will apply Homotopy perturbation method (HPM) to the nonlinear Bratu's problem, Troesch's problem, and catalytic reactions in flat particles.

Systems of non linear differential equations arise in mathematical models throughout science and engineering. When an explicit condition that a solution must satisfy is specified at one value of the independent variable, usually its lower bound, this is referred to as an initial value problem (IVP). When the conditions to be satisfied occur at more than one value of the independent variable, this is referred to as a boundary value problem (BVP). If there are two values of the independent variable at which conditions are specified, then this is a two-point boundary value problem (TPBVP). TPBVPs occur in a wide variety of problems, including the modelling of chemical reactions, heat transfer, and diffusion. They are also of interest in optimal control problems.

There are many techniques available for the numerical solution of TPBVPs for ordinary differential equations [13]. The standard techniques can be divided into two classes. Typical of this class are various shooting and multi-shooting approaches. The other class involves converting the TPBVP into a system of algebraic equations, and includes methods based on various versions of finite difference or collocation. Methods for solving TPBVPs usually require users to provide an initial guess for the unknown initial states and/or parameters.

The problem of reliably identifying all solutions of a TPBVP was apparently first addressed only recently, by [14,15]. Present a new approach that will rigorously guarantee the enclosure of all solutions to the TPBVP. In this paper we have obtained the analytical solutions of some nonlinear elliptic problems (Bratu's equation, Troesch's problem and Catalytic reactions in a flat particles) using Homotopy perturbation method.

MATHEMATICAL FORMULATION OF THE PROBLEM

Many problems in science and engineering require the computational of family of solutions of a non linear system of the form [16]:

$$G(y,\lambda)=0, \quad y=y(\lambda) \tag{1}$$

where $G:\Re^{n+1} \to \Re$ is continuously differentiable function, y represents the solution and λ is a real parameter (i.e., Reynold's number, load etc.). It is required to find a solution for some λ-interval, i.e., a path solutions, $(y(\lambda),\lambda)$. Equations of the form (1) are called nonlinear elliptic eigenvalue problems if the operator G with λ fixed is an elliptic differential operator. Fore more details about this type of operators see [17]. As a typical example of nonlinear elliptic eigenvalue problems, we consider the following problem

$$G(y,\lambda)=\Delta y+\lambda f(y), \text{ in } \Omega \tag{2}$$

$$y=0, \text{ on } \partial\Omega \tag{3}$$

where Δ is Laplacian operator in one dimension.

Equation (2) arises in many physical problems. For example, in chemical reactor theory, radiative heat transfer, combustion theory, and in modelling the expansion of the universe. The function y could be a function of several variables and the domain Ω is usually taken to be the unit interval $[0,1]$ in $\Re$, or the unit square $[0,1]\times[0,1]$ in $\Re^2$, or the unit cube $[0,1]\times[0,1]\times[0,1]$ in $\Re^3$. Equation (1) can take several forms, for example, Bratu equation is given by

$$\Delta y+\lambda e^y = 0, \text{ in } \Omega \tag{4}$$

$$y=0, \text{ on } \partial\Omega \tag{5}$$

and a reaction-diffusion problem takes the form

$$\Delta y+\lambda \exp\left(\frac{y}{1+\alpha y}\right)=0, \text{ in } \Omega \tag{6}$$

$$y=0, \text{ on } \partial\Omega \tag{7}$$

There are no bifurcation points in the two problems above; all singular points are fold points. The behaviour of the solution near the singular points has been studied numerically [17-19] and theoretically [20-23]. For both one and two-dimensional cases, the Bratu problem has exactly one fold point, whereas the three-dimensional case has infinitely many fold points.

Bratu's Equation and Its Solution

Bratu's equation [24] was first studied as a simple case of a second-order ordinary differential equation by Bratu [25]. The equation arises when deriving the temperature distribution for a reaction in an infinite vessel with planeparallel walls, and also in a simplification of a combustion reaction with a cylindrical vessel [26]. The differential equation is

$$y'' + \lambda \exp(y) = 0, \, t \in [0,1] \tag{8}$$

with boundary conditions

$$y(0) = y(1) = 0 \tag{9}$$

The analytical solution of Equations (8) and (9) using Homotopy perturbation method (See Appendix A) is

$$
\begin{aligned}
y(t) = & -\left(\frac{7+b^2}{4}\right) + \left(\frac{5+b^2}{3} - \frac{\sqrt{\lambda}b\,t}{2}\right)\cos\left(\sqrt{\lambda}t\right) \\
& + \left(\frac{1-b^2}{12}\right)\cos\left(2\sqrt{\lambda}t\right) + \left(\frac{b\sin\left(2\sqrt{\lambda}t\right)}{6}\right) \\
& + \sin\left(\sqrt{\lambda}t\right)\left\{\left(b + \frac{t\sqrt{\lambda}}{2}\right) + \left(\frac{1}{\sin\left(\sqrt{\lambda}\right)}\right)\left[\left(\frac{3+b^2}{4}\right)\right.\right. \\
& + \cos\left(2\sqrt{\lambda}t\right)\left(\frac{b^2-1}{12}\right) - \left(\frac{\sqrt{\lambda}\sin\left(\sqrt{\lambda}\right)}{2}\right) \\
& \left.\left. -\left(\frac{b\sin\left(2\sqrt{\lambda}\right)}{6}\right) + \cos\left(\sqrt{\lambda}\right)\left(\frac{b\sqrt{\lambda}}{2} - \frac{\left(b^2+2\right)}{3}\right)\right]\right\}
\end{aligned}
\tag{10}
$$

where

$$b = \left(\frac{1-\cos\left(\sqrt{\lambda}\right)}{\sin\left(\sqrt{\lambda}\right)}\right) \tag{11}$$

Reaction Diffusion Equation and Its Solution

Consider the reaction diffusion equation [16]

$$y'' + \lambda \exp\left[\frac{y}{(1+\alpha y)}\right] = 0, t \in (0,1) \tag{12}$$

with the boundary conditions

$$y(0) = y(1) = 0 \tag{13}$$

The analytical solution of Equations (12) and (13) using Homotopy perturbation method (See Appendix C) is

$$
\begin{aligned}
y(t) = &\left(\frac{\alpha(b^2+3)}{2} - 1\right) \\
&+ \left(1 - \frac{\alpha(2b^2+4)}{3} + \alpha b\sqrt{\lambda}t\right)\cos(\sqrt{\lambda}t) \\
&+ \frac{\alpha(b^2-1)\cos(2\sqrt{\lambda}t)}{6} - \left(\frac{\alpha b \sin(2\sqrt{\lambda})\,t}{3}\right) \\
&+ \sin(\sqrt{\lambda}t)\left\{(b - \alpha\sqrt{\lambda}t) + \left(\frac{\alpha}{\sin(\sqrt{\lambda})}\right)\right\}\left[\sqrt{\lambda}\sin(\sqrt{\lambda})\right. \\
&+ \frac{b\sin(2\sqrt{\lambda})(b^2+3)}{3} - \frac{(b^2-1)\cos(2\sqrt{\lambda})}{6} \\
&+ \left.\left(\frac{2b^2+4}{3} - b\sqrt{\lambda}\right)\cos(\sqrt{\lambda})\right]\}
\end{aligned}
\tag{14}
$$

where b is defined by Equation (11).

Troesch's Problem and Its Solution

Troesch's problem comes from the investigation of the confinement of a plasma column under radiation pressure. The problem was first described and solved by Weibel [27]. It has become a widely used test problem, and has been solved many times, including in analytical closed form [28] by using a shooting method [29], by using a Laplace transform decomposition technique [30] and most recently by using a modified Homotopy perturbation technique [31]. The differential equation is

$$y'' = \lambda \sinh(\lambda y),\ t \in [0,1] \tag{15}$$

with the boundary conditions

$$y(0) = 0 \text{ and } y(1) = 1 \tag{16}$$

The known analytical, closed form solution [28] of Equations (15) and (16) is given by

$$y(t) = \frac{2}{\lambda} \sinh^{-1}\left[\frac{y'(0)}{2} sc\left(\lambda t, 1 - \frac{1}{4}(y'(0))^2\right)\right]$$

(17)

where $y'(0) = 2(1-m)^{\frac{1}{2}}$ is the derivative at $t = 0$ and the constant m is the solution to the equation

$$\frac{\sinh\left(\frac{\lambda}{2}\right)}{(1-m)} = sc(\lambda, m)$$

(18)

We have obtained the analytical solution of Equations (15) and (16) using Homotopy perturbation method (See Appendix F) is

$$y(t) = \left(\frac{\sinh(\lambda t)}{\sinh(\lambda)}\right)$$
$$+ \left(\frac{\lambda^3}{48\sinh^3(\lambda)}\right)\left\{\left(\frac{\sinh(\lambda t)}{\sinh(\lambda)}\right)\left(3\cosh(\lambda) - \frac{\sinh(3\lambda)}{4}\right)\right.$$
$$\left. + \left(\frac{\sinh(3\lambda t)}{4\lambda}\right) - 3t\cosh(\lambda t)\right\}$$

(19)

Catalytic Reactions in a Flat Particle and Its Solution

This example arises in a study of heat and mass transfer for a catalytic reaction within a porous catalyst flat particle [32]. The differential equation is the direct result of a material and energy balance. Assuming a flat geometry for the particle and that conductive heat transfer is negligible compared to convective heat transfer yields the differential equation.

$$y'' = \lambda y \exp\left[\frac{\gamma\beta(1-y)}{1+\beta(1-y)}\right], t \in [0,1]$$

(20)

with boundary conditions

$$y'(0) = 0 \text{ and } y(1) = 1$$

(21)

The analytical solution of the Equations (20) and (21) using Homotopy perturbation method [33-41] (See Appendix H) is

$$y(t) = \left(1 + \frac{\lambda\beta\gamma(\cosh(2k)-3)}{6k^2(1+\beta)^2\cosh^2(k)}\right)\left(\frac{\cosh(kt)}{\cosh(k)}\right)$$
$$+ \left(\frac{\lambda\gamma\beta(3-\cosh(2kt))}{6(1+\beta)^2\cosh^2(k)}\right)$$

(22)

where

$$k = \sqrt{\lambda + \frac{\lambda\beta\gamma}{(1+\beta)}}$$

(23)

NUMERICAL SIMULATION

The non-linear equations [Equations (3), (7), (10) and (15)] for the given boundary conditions are solved by numerically. The function pdex4, in Matlab software is used to solve two-point boundary value problems (BVPs) for ordinary differential equations given in Appendix B, Appendix D, Appendix E, Appendix G, Appendix I, Appendix J and Appendix K. The numerical results are also compared with the obtained analytical expressions [Equations (5), (6), (9), (14), (17) and (18)] for all values of parameters λ, α, β and γ.

RESULTS AND DISCUSSION

Figure 1 represents the dimensionless concentration $y(t)$ versus the dimensionless distance t for different values of the dimensionless parameter λ. From this figure, it is evident that the values of the dimensionless concentration $y(t)$ increases when dimensionless parameter λ increases. Figures 2(a)-(d) show the concentration $y(t)$ versus dimensionless distance t for various values of dimensionless parameters α and λ.

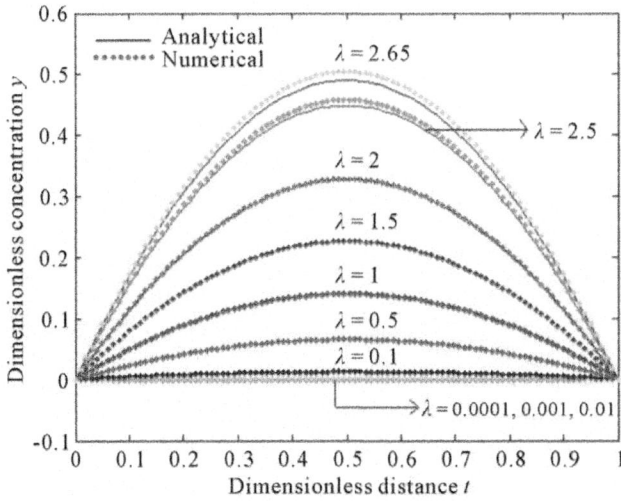

Figure 1: The curve is plotted for the influence of λ on the dimensionless on concentration y(t) versus the dimensionless distance t obtained from the Equations (10) and (11).

From these figures, it is obvious that the values of the dimensionless concentration $y(t)$ increases when dimensionless parameters λ increases for the fixed values of α. From the Figures 3(a) and (b), it is clear that the concentration $y(t)$ decreases for the different values of the dimensionless parameter α, for the various values of λ. The dimensionless concentration $y(t)$ versus the dimensionless distance t for different values of dimensionless parameter λ is plotted in Figure 4.

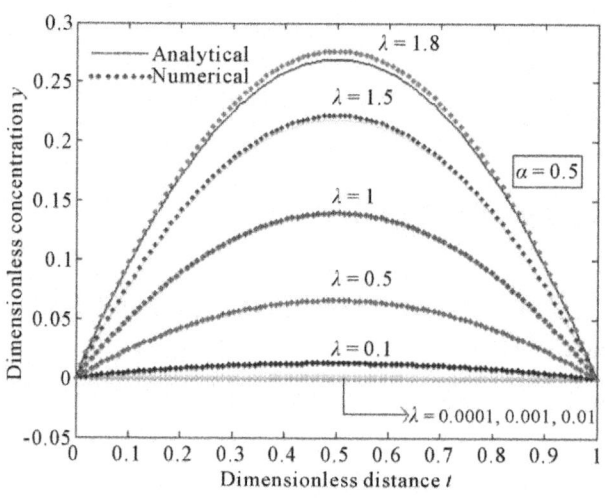

(a)

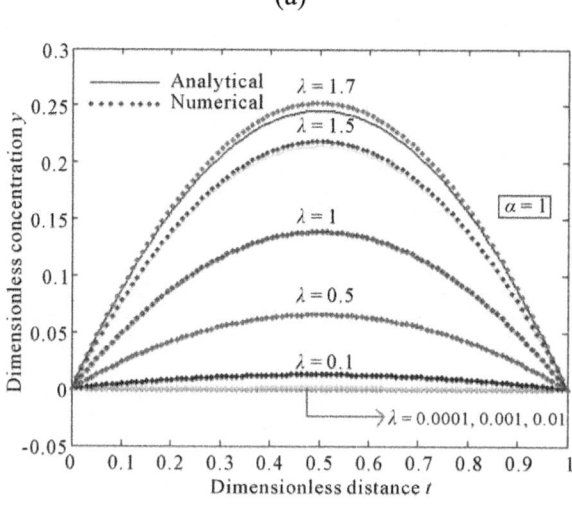

(b)

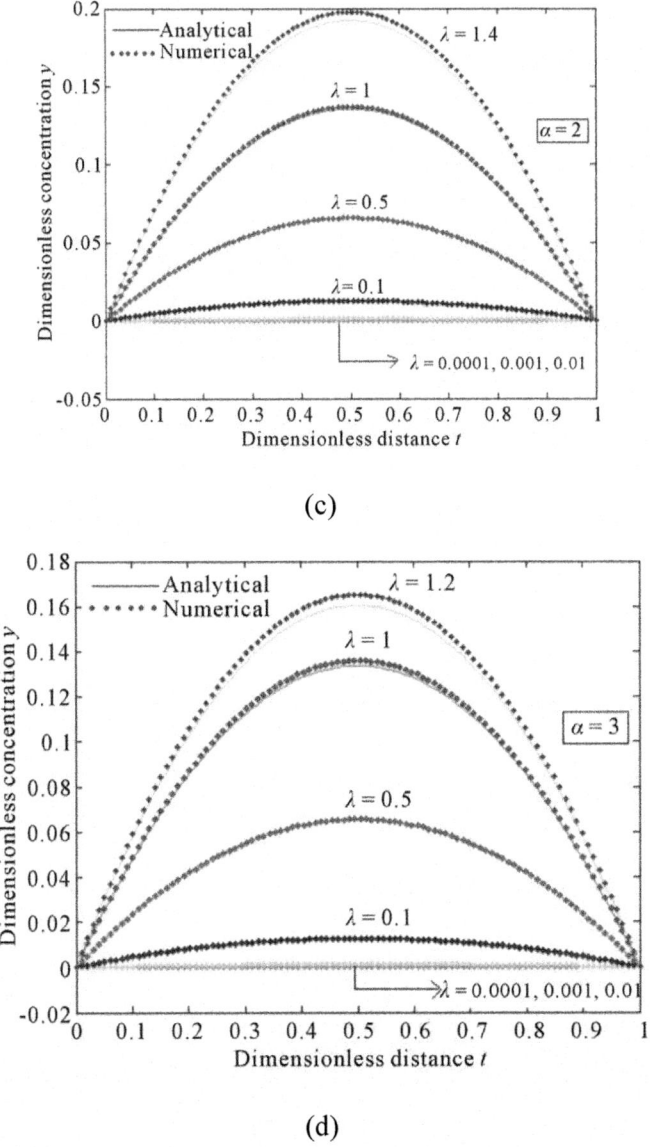

(c)

(d)

Figure 2: Influence of λ on the dimensionless concentration $y(t)$ obtained from the Equation (14). The curve is plotted, when (a) $\alpha = 0.5$; (b) $\alpha = 1$; (c) $\alpha = 2$; (d) $\alpha = 3$.

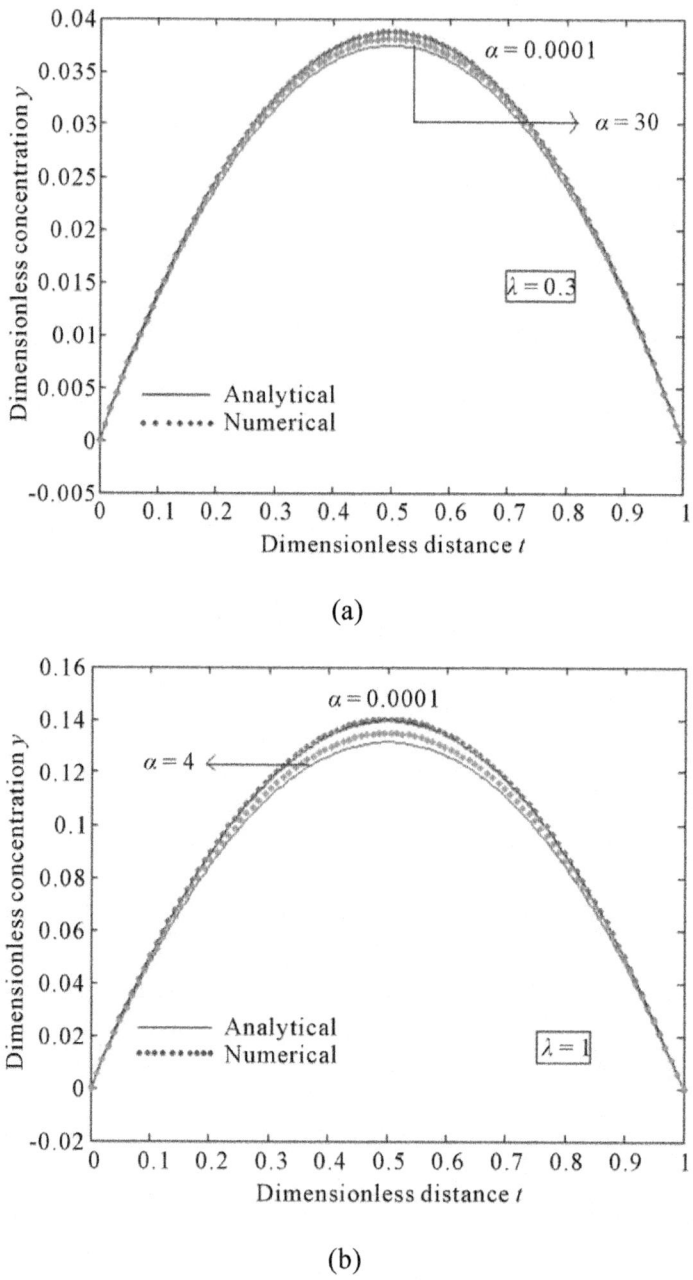

(a)

(b)

Figure 3: Influence of α on the dimensionless concentration y(t) obtained from the Equation (14). The curve is plotted, when (a) λ = 0.3; (b) λ = 1.

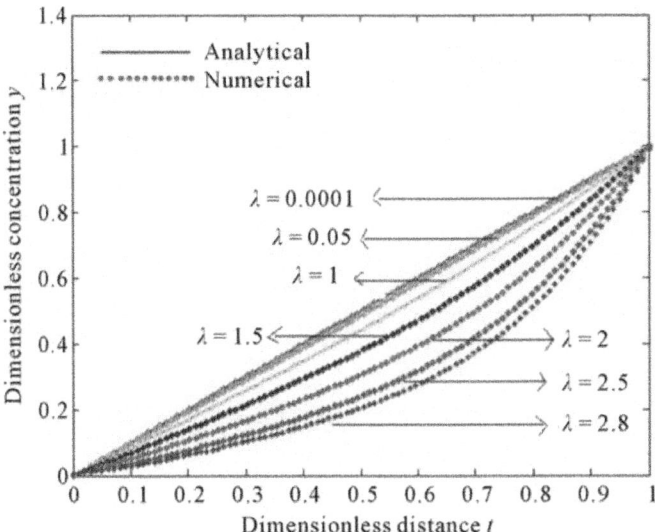

Figure 4: The curve is plotted for the influence of λ on the dimensionless concentration y(t) versus the dimensionless distance t from the Equation (19).

From this figure, it shows that the concentration $y(t)$ decreases for the various values of λ. Figures 5(a)-(d) shows the dimensionless concentration $y(t)$ in the reactor versus the dimensionless distance down the reactor t. From these figures it is clear that the concentration $y(t)$ decreases for the fixed values of α and γ for the different values of λ.

Figures 6 and 7 shows the dimensionless concentration $y(t)$ versus the dimensionless distance t. From these figures it is clear that the concentration $y(t)$ decreases for the fixed values of λ and γ for the different values of α.

CONCLUSION

The steady state non-linear reaction-diffusion equation has been solved analytically and numerically. The dimensionless concentrations $y(t)$ in the reactor at the position t are derived by using the HPM. The primary result of this work is simple approximate calculations of concentration for all values of dimensionless parameters α, β, γ and λ. The HPM is an extremely simple method and it is also a promising method to solve other non-linear equations. This method can be easily extended to find the solution of all other non-linear equations.

ACKNOWLEDGEMENTS

This work was supported by the University Grants Commission (F. No. 39-58/2010(SR)), New Delhi, India. The authors are thankful to Mr. M. S. Meenakshisundaram, The Secretary, Dr. R. Murali, The Principal and Dr.

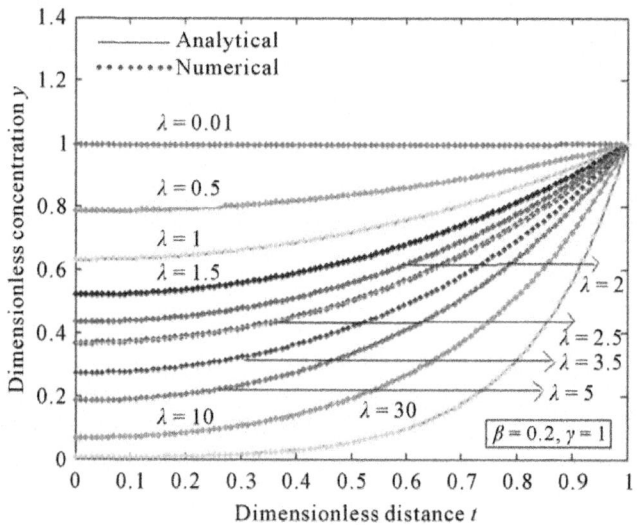

(a)

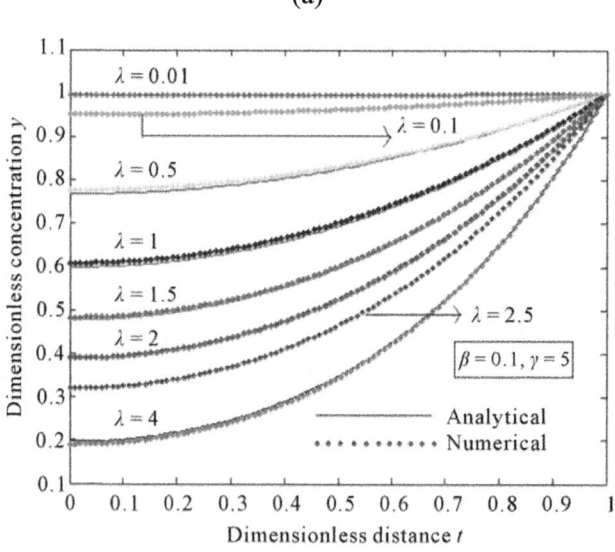

(b)

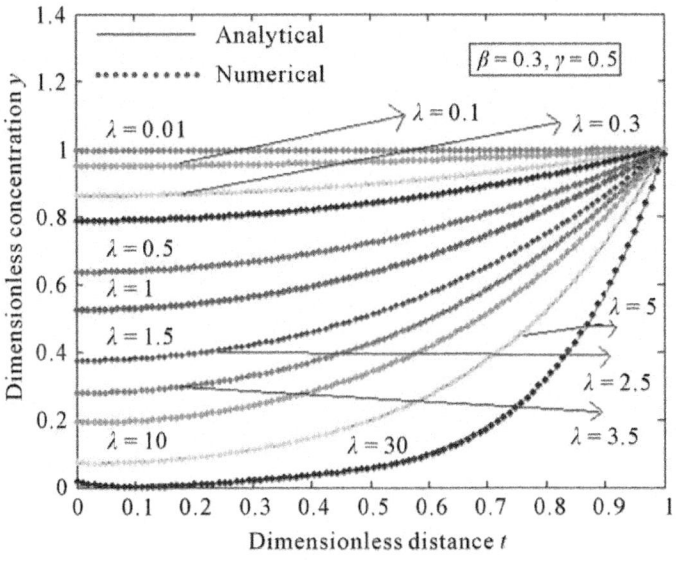

(c)

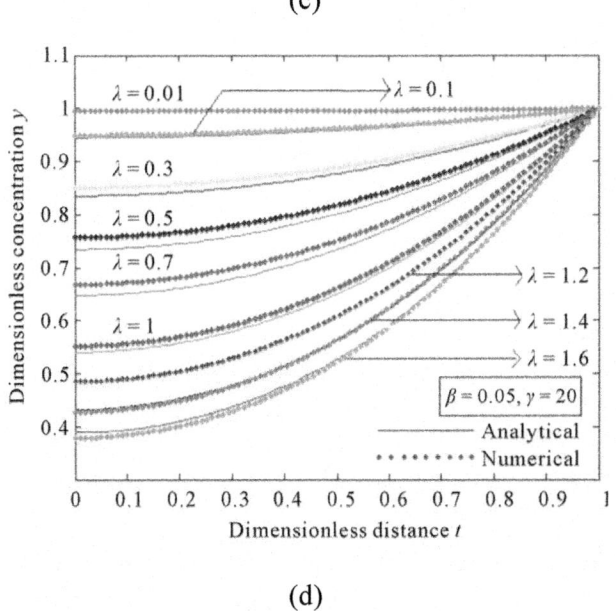

(d)

Figure 5: The curve is plotted for the influence of λ on the dimensionless concentrattion y versus the dimensionless distance down the reactor t obtained from Equations (22) and (23), when (a) $\beta = 0.2$, $\gamma = 1$; (b) $\beta = 0.1$, $\gamma = 5$; (c) $\beta = 0.05$, $\gamma = 20$; (d) $\beta = 0.3$, $\gamma = 0.5$.

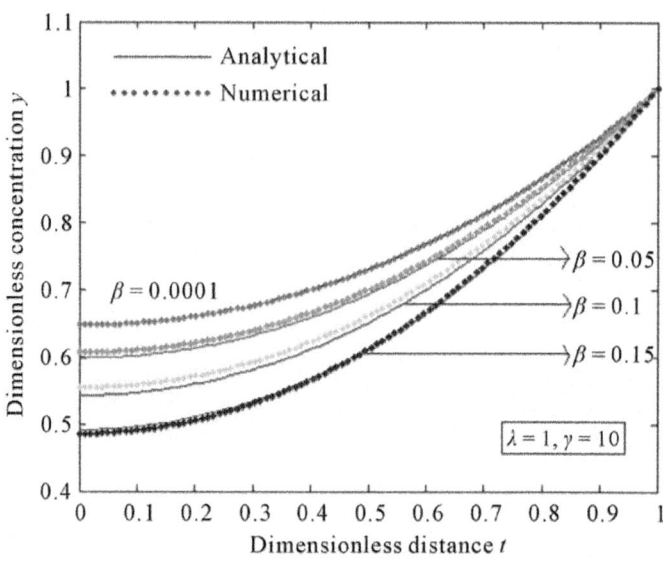

(a)

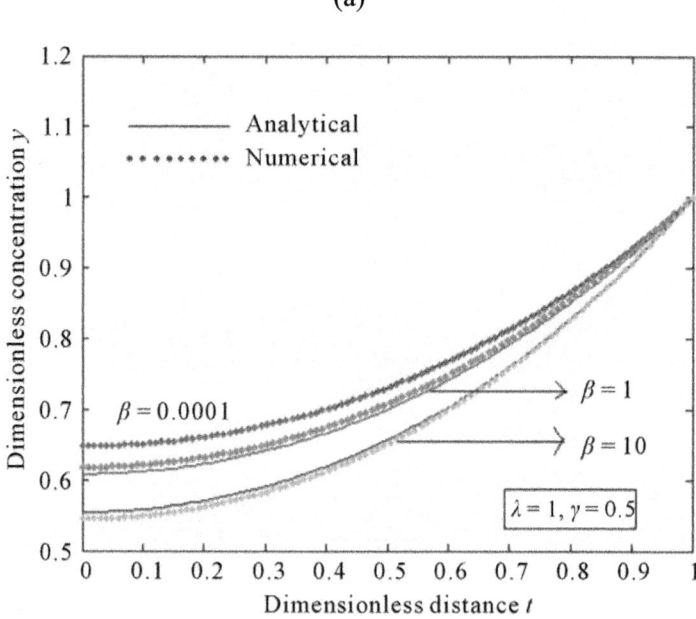

(b)

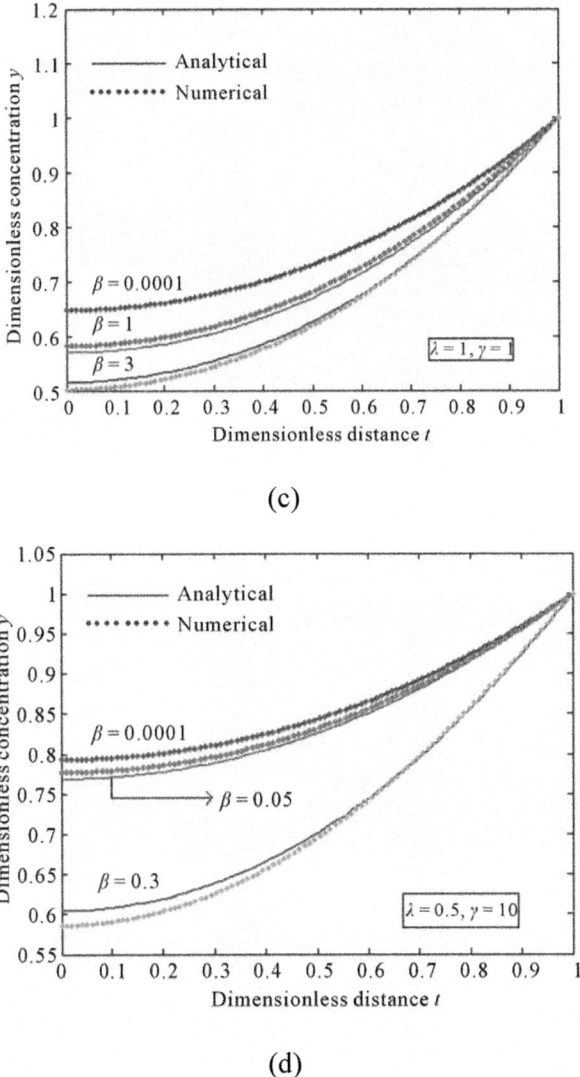

(c)

(d)

Figure 6: The curve is plotted for the influence of β on the dimensionless concentration y versus the dimensionless distance down the reactor t obtained from Equations (22) and (23), when (a) λ = 1, γ = 10; (b) λ = 1, γ = 0.5; (c) λ = 1, γ = 1; (d) λ = 0.5, γ = 10.

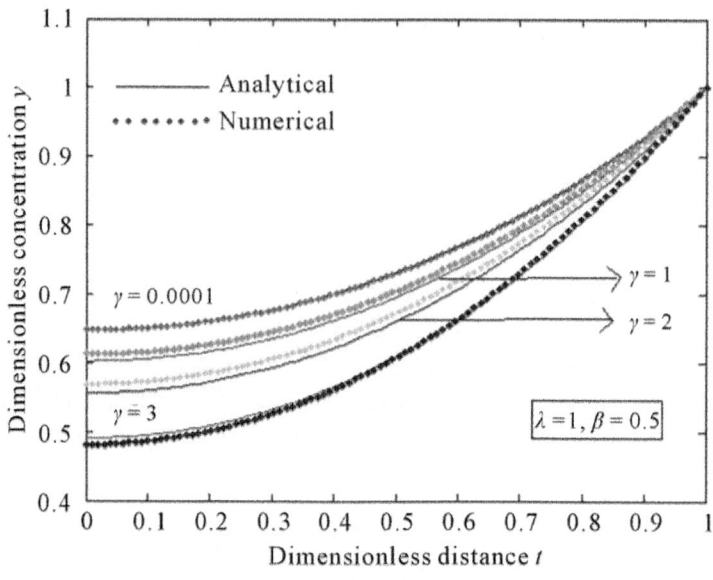

(a)

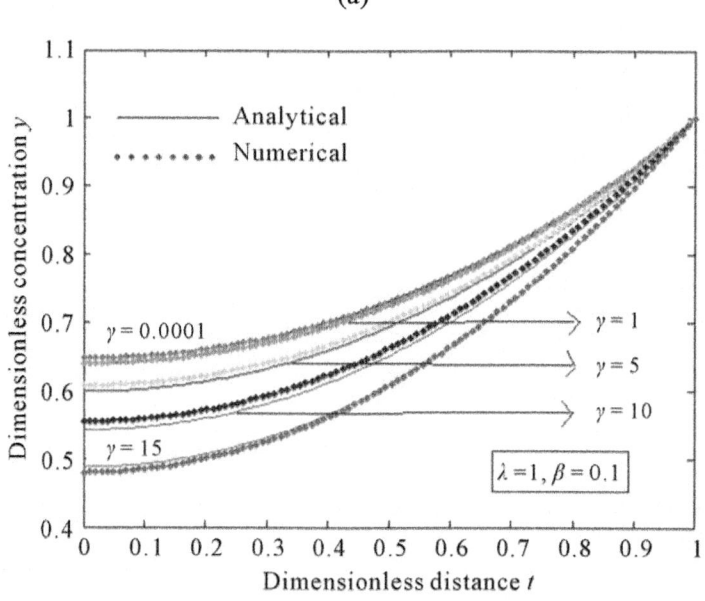

(b)

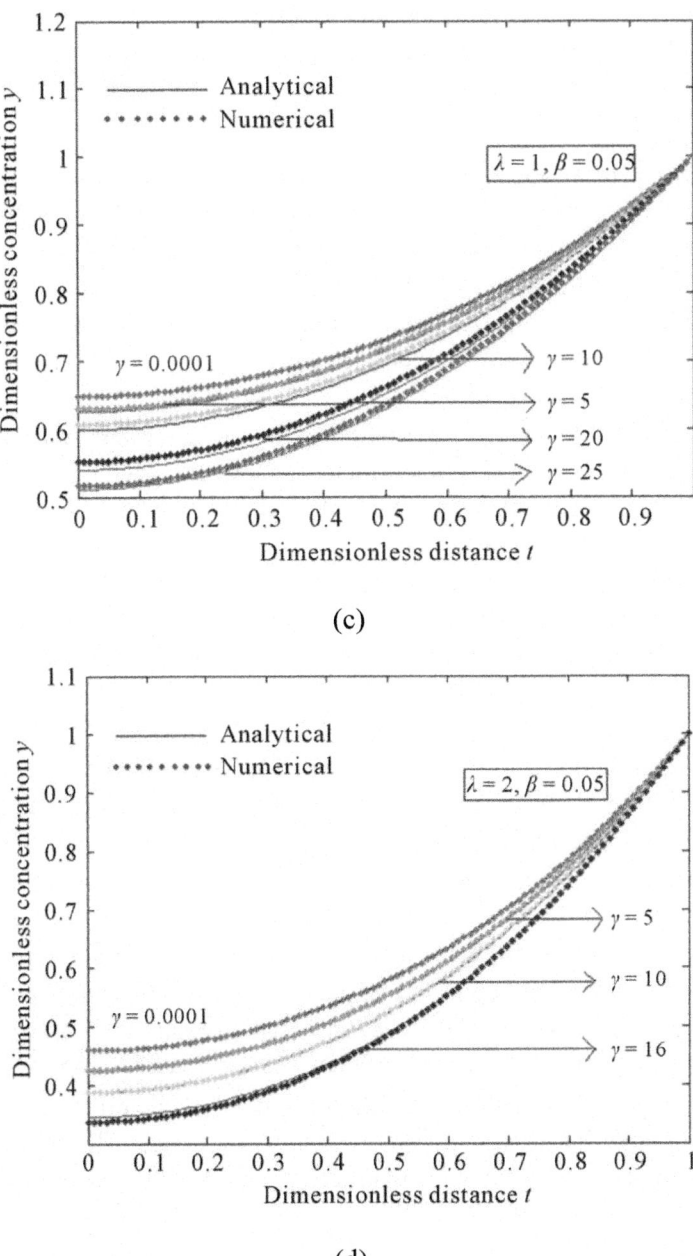

(c)

(d)

Figure 7: The curve is plotted for the influence of γ on the dimensionless concentration y versus the dimensionless distance down the reactor t obtained from Equations (22) and (23), when (a) $\lambda = 1$, $\beta = 0.5$; (b) $\lambda = 1$, $\beta = 0.1$; (c) $\lambda = 1$, $\beta = 0.05$; (d) $\lambda = 2$, $\beta = 0.05$.

L. Rajendran, Assistant Professor, Department of Mathematics, The Madura College, Madurai for their encouragement.

REFERENCES

1.   Y. Lin, J. A. Enszer and M. Stadtherr, "Enclosing All Solutions of Two Point Boundary Value Problems for ODEs," Computers and Chemical Engineering, Vol. 32, 2008, pp. 1714-1725. doi:10.1016/j.compchemeng.2007.08.013

2.   J. Newman and W. Tiedmann, "Porous-Electrode Theory with Battery Applications," AIChE Journal, Vol. 21, No. 1, 1975, pp. 25-41. doi:10.1002/aic.690210103

3.   S. H. Mirmoradi, I. Hosseinpour, S. Ghanbarpour and A. Barari, "Application of an Approximate Analytical Method to Nonlinear Troesch's Problem," Applied Mathematical Sciences, Vol. 3, No. 32, 2009, pp. 1579-1585.

4.   Barari, A. R. Ghotbi, F. Farrokhzad and D. D. Ganji, "Variational Iteration Method and Homotopy-Perturbation Method for Solving Different Types of Wave Equations," Journal of Applied Sciences, Vol. 8, 2008, pp. 120-126. doi:10.3923/jas.2008.120.126

5.   R. Abdoul, A. Ghotbi, A. Barari and D. D. Ganji, "Solving Ratio-Dependent Redator-Prey System with Constant Effort Harvesting Using Homotopy Perturbation Method," Journal of Mathematical Problems in Engineering, 2008, Article ID: 945420.

6.   Barari, A. Janalizadeh and D. D. Ganji, "Application of Homotopy Perturbation Method to Zakharov-Kuznetsov Equation," Journal of Physics, Vol. 96, 2008, pp. 1-8. doi:10.1088/1742-6596/96/1/012082

7.   Barari, D. D. Ganji and M. J. Hosseini, "HomotopyPerturbation Method for a Nonlinear Cerebral ReactionDiffusion Equation," Arab Journal of Mathematics and Mathematical Sciences, Vol. 1, 2007, pp. 1-9.

8.   Barari, M. Omidvar, A. R. Ghotbi and D. D. Ganji, "Application of Homotopy-Perturbation Method and Variational Iteration Method to Nonlinear Oscillator Differential Equations," Acta Applicanda Mathematicae, Vol. 104, 2008, pp. 161-171. doi:10.1007/s10440-008-9248-9

9.   Barari, M. Omidvar, D. D. Ganji and Abbas Tahmasebi poor, "An Approximate Solution for Boundary Value Problems in Structural Engineering and Fluid Mechanics," Journal of Mathematical Problems in Engineering, 2008, Article ID: 394103.

10. L. N. Zhang and J. H. He, "Homotopy-Perturbation Method for the Solution of the Electrostatic Potential Differential Equation," Mathematical Problems in Engineering, 2006, Article ID: 83878. doi:10.1155/MPE/2006/83878

11. J. H. He, "New Interpretation of Homotopy-Perturbation Method," International Journal of Modern Physics B, Vol. 20, No. 18, 2006, pp. 256-2568. doi:10.1142/S0217979206034819

12. J.-H. He, "Homotopy-Perturbation Technique," Computer Methods in Applied Mechanics and Engineering, Vol. 178, No. 3-4, 1999, pp. 257-262. doi:10.1016/S0045-7825(99)00018-3

13. J.-H. He, "Homotopy-Perturbation Method: A New NonLinear Analytical Technique," Applied Mathematics and Computation, Vol. 135, No. 1, 2003, pp. 73-79. doi:10.1016/S0096-3003(01)00312-5

14. Y. Chen, "Dynamic System Optimization," Ph.D. Thesis, University of California, Los Angeles, 2006.

15. Y. Chen and V. Manousiouthakis, "Identification of All Solutions of TPBV Problems," AIChE Annual Meeting, Cincinnati, Dover, New York, 2005.

16. M. I. Syam and E. M. Allon, "On the Computations of Fold Points for Non Linear Elliptic Eigen Value Problems," International Journal of Open Problems in Computer Science and Mathematics, Vol. 4, No. 1, 2011.

17. J. P. Abbott, "An Efficient Algorithm for the Determination of Certain Bifurcation Point," Journal of Computational and Applied Mathematics, Vol. 4, No. 1, 1978, pp. 19-27. doi:10.1016/0771-050X(78)90015-3

18. H. Amann, "Fixed Point Equations and Nonlinear Eigenvalue Problems in Ordered Banach Space," SIAM Review, Vol. 8, 1978, pp. 19-27.

19. H. B. Keller and D. S. Choen, "Some Positone Problems Suggested by Nonlinear Heat Generation," Journal of Mathematics and Mechanics, Vol. 16, 1967, pp. 1361- 1376.

20. U. M. Assher, R. M. Matthij and R. D. Russell, "Numerical Solution of Boundary Value Problems for Ordinary Differential Equations," Society for Industrial and Applied Mathematics, Philadelphia, 1995.

21. G. Moore and A. Spence, "The Calculation of Turning Points of Nonlinear Equations," SIAM Journal on Numerical Analysis, Vol. 17, No. 4, 1980, pp. 567-576. doi:10.1137/0717048

22. E. Deeba, S. A. Khuri and S. Xie, "An Algorithm for Solving Boundary Value Problems," Journal of Computational Physics, Vol. 159, No. 2, 2000, pp. 125-138. doi:10.1006/jcph.2000.6452

23.   J. B. Rosen, "Approximate Solution and Error Bounds for Quasilinear Elliptic Boundary Value Problems," SIAM Journal on Numerical Analysis, Vol. 7, No. 1, 1970, pp. 80-103. doi:10.1137/0707004

24.   H. Davis, "Introduction to Nonlinear Differential and Integral Equations," Dover, New York, 1962.

25.   G. Bratu, "Sur Certaines Equations Integrals Non Lineares," Comptes Rendus, Vol. 150, 1910, pp. 896-899.

26.   D. A. Frank-Kamenetskii, "Diffusion and Heat Transfer in Chemical Kinetics," Plenum Press, New York, 1969.

27.   E. S. Weibel, "Confident of a Plasma Column by Radiation Pressure," In: R. K. M. Landshoff, Ed., The Plasma in a Magnetic Field, Stanford University Press, Stanford, 1958, pp. 60-76.

28.   S. M. Roberts and J. Shipman, "On the Closed Form Solution of Troesch's Problem," Journal of Computational Physics, Vol. 21, No. 3, 1976, pp. 291-304. doi:10.1016/0021-9991(76)90026-7

29.   A. Troesch, "A Simple Approach to a Sensitive TwoPoint Boundary Value Problem," Journal of Computational Physics, Vol. 21, No. 3, 1976, pp. 279-290. doi:10.1016/0021-9991(76)90025-5

30.   S. A. Khuri, "A Numerical Algorithm for Solving Troesch's Problem," International Journal of Computer Mathematics, Vol. 80, No. 4, 2003, pp. 493-498. doi:10.1080/0020716022000009228

31.   X. Feng, L. Mei and G. He, "An Efficient Algorithm for Solving Troesch's Problem," Applied Mathematics and Computation, Vol. 189, No. 1, 2007, pp. 500-507. doi:10.1016/j.amc.2006.11.161

32.   V. Hlavacek, M. Marek and M. Kubicek, "Modelling of Chemical Reactors, X: Multiple Solutions of Enthalpy and Mass Balance for a Catalytic Reaction within a Porous Catalyst Particle," Chemical Engineering Science, Vol. 23, No. 9, 1968, pp. 1083-1097. doi:10.1016/0009-2509(68)87093-9

33.   J. H. He, "Homotopy Perturbation Technique," Computer Methods in Applied Mechanics and Engineering, Vol. 178, No. 3-4, 1999, pp. 257-262. doi:10.1016/S0045-7825(99)00018-3

34.   J. H. He, "Homotopy Perturbation Method: A New Nonlinear Analytical Technique," Applied Mathematics and Computation, Vol. 135, No. 1, 2003, pp. 73-79. doi:10.1016/S0096-3003(01)00312-5

35.   J. H. He, "A Simple Perturbation Approach to Blasius Equation," Applied Mathematics and Computation, Vol. 140, No. 2-3, 2003, pp. 217-222. doi:10.1016/S0096-3003(02)00189-3

36. P. D. Ariel, "Alternative Approaches to Construction of Homotopy-Perturbation Algorithms," Nonlinear Science Letters A, Vol. 1, 2010, pp. 43-52.

37. V. Ananthaswmy, A. Eswari and L. Rajendran, "Analytical Solution of System of Nonlinear Reaction-Diffusion Equations in a Thin Membrane: Homotopy-Perturbation Approach," Journal of Physical Chmistry, Vol. 5, No. 2, 2010.

38. S. Loghambal and L. Rajendran, "Mathematical Modeling of Diffusion and Kinetics of Amperometric Immobilized Enzyme Electrodes," Electrochim Acta, Vol. 55, No. 18, 2010, pp. 5230-5238. doi:10.1016/j.electacta.2010.04.050

39. Meena and L. Rajendran, "Mathematical Modeling of Amperometric and Potentiometric Biosensors and System of Non-Linear Equations, Homotopy-Perturbation Approach," Journal of Electroanalytical Chemistry, Vol. 644, No. 1, 2010, pp. 50-59. doi:10.1016/j.jelechem.2010.03.027

40. S. Anitha, A. Subbiah, S. Subramaniam and L. Rajendran, "Analytical Solution of Amperometric Enzymatic Reactions Based on Homotopy-Perturbation Method," Electrochimica Acta, Vol. 56, No. 9, 2011, pp. 3345-3352. doi:10.1016/j.electacta.2011.01.014

41. V. Ananthaswamy and L. Rajendran, "Analytical Solution of Two-Point Non Linear Boundary Value Problems in a Porous Catalyst Particles," International Journal of Mathematical Archive, Vol. 3, No. 3, 2012, pp. 810-821.

Appendix A: Solution of Bratu's Equation Using HPM

In this Appendix, we indicate how the Equation (10) is derived. When y is small, Equation (8) is reduces to

$$\frac{d^2 y}{dt^2} + \lambda \left[1 + y + \frac{y^2}{2} \right] = 0$$

(A1)

We construct the Homotopy for the Equation (A1) is as follows:

$$(1-p)\left[\frac{d^2 y}{dt^2} + \lambda y + \lambda\right] + p\left[\frac{d^2 y}{dt^2} + \lambda y + \lambda + \frac{\lambda y^2}{2}\right] = 0$$

(A2)

The analytical solution of Equation (8) with Equation (9) is

$$y = y_0 p y_1 + p^2 y_2 + \cdots$$

(A3)

Substituting the Equation (A3) into an Equation (A2) we get

$$(1-p)\left[\frac{d^2\left(y_0+py_1+p^2y_2+\cdots\right)}{dt^2}\right.$$

$$\left.+\lambda\left(y_0+py_1+p^2y_2+\cdots\right)+\lambda\right]$$

$$+p\left[\frac{d^2\left(y_0+py_1+p^2y_2+\cdots\right)}{dt^2}\right.$$

$$+\lambda\left(y_0+py_1+p^2y_2+\cdots\right)+\lambda$$

$$\left.+\frac{\lambda\left(y_0+py_1+p^2y_2+\cdots\right)^2}{2}\right]=0$$

$$(A4)$$

Comparing the coefficients of like powers of p in Equation (A4) we get

$$p^0:\frac{d^2y_0}{dt^2}+\lambda y_0+\lambda=0 \tag{A5}$$

$$p^1:\frac{d^2y_1}{dt^2}+\lambda y_1+\frac{\lambda y_0^2}{2}=0 \tag{A6}$$

The initial approximations are as follows

$$y_0(0)=0,\ y_0(1)=0, \tag{A7}$$

$$i=1,2,3,\cdots\ y_i(0)=y_i(1)=0, \tag{A8}$$

Solving the Equation (A5) and the Equation (A6) and using the boundary conditions Equation (A7) and the Equation (A8) we obtain the following results:

$$y_0=\left(\cos\left(\sqrt{\lambda}t\right)-1\right)+b\sin\left(\sqrt{\lambda}t\right) \tag{A9}$$

$$
y_1 = \left[\left[\left(\frac{b^2+2}{3} \right) - \frac{\sqrt{\lambda}bt}{2} \right] \cos\left(\sqrt{\lambda}t \right) + \left(\frac{1-b^2}{12} \right) \cos\left(2\sqrt{\lambda}t \right) \right.
$$

$$
+ \left(\frac{b\sin(2\sqrt{\lambda}t)}{6} \right) - \left(\frac{b^2+3}{4} \right) + \sin(\sqrt{\lambda}t)\left\{ b + \frac{\sqrt{\lambda}t}{2} \right.
$$

$$
+ \frac{1}{\sin\left(\sqrt{\lambda} \right)} \left[\left(\frac{b^2+3}{4} \right) - \left(\frac{b^2-1}{12} \right) \cos\left(2\sqrt{\lambda} \right) \right.
$$

$$
- \left(\frac{\lambda\sin\left(\sqrt{\lambda} \right)}{2} \right) - \left(\frac{b\sin\left(2\sqrt{\lambda} \right)}{6} \right)
$$

$$
\left. \left. \left. + \cos\left(\sqrt{\lambda} \right) \left[\left(\frac{b\sqrt{\lambda}}{2} \right) - \left(\frac{b^2+2}{3} \right) \right] \right] \right\} \right.
$$

$$
\tag{A10}
$$

where b is defined in Equation (9). According to the HPM, we can conclude that

$$
y = \lim_{p \to 1} y(t) = y_0 + y_1
\tag{A11}
$$

After putting the Equation (A9) and Equation (A10) into an Equation (A11) we obtain the solution in the text.

Appendix B: Matlab Program Is to Find the Numerical Solution of the Non Linear Differential Equations (8) and (9)

```
function pdex4 m = 0;
x = linspace(0,1);
t=linspace(0,10000);
sol= pdepe(m,@pdex4pde,@pdex4ic,@pdex4bc,x,t);
u1 = sol(:,:,1);
figure plot(x,u1(end,:))
title('u1(x,t)')
xlabel('Distance x')
ylabel('u1(x,2)')
%-----------------------------------------------------------------
function [c,f,s] = pdex4pde(x,t,u,DuDx)
c = 1;
f = DuDx;
lamda=2;
```

F =lamda*exp(u)

s = F;

%---

function u0 = pdex4ic(x); %create a initial conditions u0 = 1;

%---

function[pl,ql,pr,qr]=pdex4bc(xl,ul,xr,ur,t) %create a boundary conditions pl = ul;

ql = 0;

pr = ur-0;

qr = 0;

Appendix C: Solution of Reaction Diffusion Equation Using HPM

In this Appendix, we indicate how Equation (14) is derived. When $\overline{\frac{y}{1+\alpha y}}$ is small, Equation (12) is reduces to

$$\frac{d^2 y}{dt^2} + \lambda \left[1 + y - \alpha y^2\right] = 0$$

(C1)

We construct the Homotopy for Equation (C1) is as follows:

$$(1-p)\left[\frac{d^2 y}{dt^2} + \lambda y + \lambda\right] + p\left[\frac{d^2 y}{dt^2} + \lambda y + \lambda - \lambda \alpha y^2\right] = 0$$

(C2)

The analytical solution of Equation (12) with Equation (13) is

$$y = y_0 + p y_1 + p^2 y_2 + \cdots$$

(C3)

Substituting Equation (C3) into an Equation (C2) we get

$$(1-p)\left[\frac{d^2\left(y_0 + p y_1 + p^2 y_2 + \cdots\right)}{dt^2}\right.$$

$$+\lambda\left(y_0 + p y_1 + p^2 y_2 + \cdots\right) + \lambda\Bigg]$$

$$-p\left[\frac{d^2\left(y_0 + p y_1 + p^2 y_2 + \cdots\right)}{dt^2}\right.$$

$$+\lambda\left(y_0 + p y_1 + p^2 y_2 + \cdots\right)$$

$$-\lambda \alpha \left(y_0 + p y_1 + p^2 y_2 + \cdots\right)^2 + \lambda\Bigg] = 0$$

(C4)

Comparing the coefficients of like powers of p in Equation (C4) we get

$$p^0 : \frac{d^2 y_0}{dt^2} + \lambda y_0 + \lambda = 0 \tag{C5}$$

$$p^1 : \frac{d^2 y_1}{dt^2} + \lambda y_1 - \lambda \alpha y_0^2 = 0 \tag{C6}$$

The initial approximations are as follows:

$$y_0(0) = 0, \; y_0(1) = 0, \tag{C7}$$

$$y_i(0) = y_i(1) = 0, \; i = 1, 2, 3 \cdots \tag{C8}$$

Solving the Equations (C5) and (C6) and using the boundary conditions Equation (C7) and the Equation (C8) we obtain the following results:

$$y_0 = \left(\cos\left(\sqrt{\lambda}t\right) - 1 \right) + b \sin\left(\sqrt{\lambda}t\right) \tag{C9}$$

$$y_1 = \left[\frac{\alpha(b^2 + 3)}{2} \right] + \left(\frac{1 - b^2}{12} \right) \cos\left(2\sqrt{\lambda}t\right)$$

$$+ \left(\frac{b\sin(2\sqrt{\lambda})}{3} \right) - \left(\frac{b^2 + 3}{2} \right) + \alpha \sin\left(\sqrt{\lambda}t\right) \left[-t\sqrt{\lambda} \right.$$

$$+ \frac{1}{\sin(\sqrt{\lambda})} \left[\left(\sqrt{\lambda} \sin\left(\sqrt{\lambda}\right) \right) - \left(\frac{b^2 - 1}{12} \right) \cos\left(2\sqrt{\lambda}\right) \right.$$

$$- \left(\frac{\lambda \sin(\sqrt{\lambda})}{2} \right) - \left(\frac{b\sin(2\sqrt{\lambda})}{6} \right)$$

$$\left. \left. + \cos\left(\sqrt{\lambda}\right) \left[\left(\frac{b\sqrt{\lambda}}{2} \right) - \left(\frac{b^2 + 2}{3} \right) \right] \right] \right] \tag{C10}$$

where b is defined in the text Equation (6). According to the HPM, we can conclude that

$$y = \lim_{p \to 1} y(t) = y_0 + y_1 \tag{C11}$$

After putting Equation (C9) and Equation (C10) into an Equation (C11) we obtain the solution in the text.

Appendix D: Matlab Program Is to Find the Numerical Solution of the Non-Linear Differential Equations (12) and (13)

function pdex4 m = 0;

```
x = linspace(0,1);
t=linspace(0,10000);
sol= pdepe(m,@pdex4pde,@pdex4ic,@pdex4bc,x,t);
u1 = sol(:,:,1);
figure plot(x,u1(end,:))
title('u1(x,t)')
xlabel('Distance x')
ylabel('u1(x,2)')
%-----------------------------------------------------------------
function [c,f,s] = pdex4pde(x,t,u,DuDx)
c = 1;
f = DuDx;
lamda=1.5;
alpha=0.5;
F =lamda*exp(u/(1+(alpha*u)));
s = F;
%-----------------------------------------------------------------
function u0 = pdex4ic(x); %create a initial conditions u0 = 1;
%-----------------------------------------------------------------
function[pl,ql,pr,qr]=pdex4bc(xl,ul,xr,ur,t) %create a boundary conditions pl
= ul;
ql = 0;
pr = ur-0;
qr = 0;
```

Appendix E: Matlab Program Is to Find the Numerical Solution of the Non Linear Differential Equations (12) and (13)

```
function pdex4 m = 0;
x = linspace(0,1);
t=linspace(0,10000);
sol= pdepe(m,@pdex4pde,@pdex4ic,@pdex4bc,x,t);
u1 = sol(:,:,1);
figure plot(x,u1(end,J)
```

```
title('u1(x,t)')
xlabel('Distance x')
ylabel('u1(x,2)')
%-------------------------------------------------------------------
function [c,f,s] = pdex4pde(x,t,u,DuDx)
c = 1;
f = DuDx;
lamda=0.3;
alpha=30;
F =lamda*exp(u/(1+(alpha*u)));
s = F;
%-------------------------------------------------------------------
function u0 = pdex4ic(x); %create a initial conditions u0 = 1;
%-------------------------------------------------------------------
function[pl,ql,pr,qr]=pdex4bc(xl,ul,xr,ur,t) %create a boundary conditions pl
= ul;
ql = 0;
pr = ur-0;
qr = 0;
```

Appendix F: Solution of Troesch's Problem Using HPM

In this Appendix, we indicate how the Equation (19) is derived.

When λy is small, Equation (15) is reduces to

$$\frac{d^2 y}{dt^2} - \lambda \left[\lambda y + \frac{\lambda^3 y^3}{6} \right] = 0 \qquad \text{(E1)}$$

We construct the Homotopy for the Equation (E1) is as follows:

$$(1-p)\left[\frac{d^2 y}{dt^2} - \lambda^2 y \right] + p\left[\frac{d^2 y}{dt^2} - \lambda^2 y - \frac{\lambda^4 y^3}{6} \right] = 0 \qquad \text{(E2)}$$

The analytical solution of Equation (15) with Equation (16) is

$$y = y_0 + py_1 + p^2 y_2 + \cdots$$

(E3)

Substituting the Equation (E3) into an Equation (E2) we get

$$(1-p)\left[\frac{d^2\left(y_0 + py_1 + p^2 y_2 + \cdots\right)}{dt^2}\right.$$

$$\left. -\lambda^2\left(y_0 + py_1 + p^2 y_2 + \cdots\right)\right]$$

$$+p\left[\frac{d^2\left(y_0 + py_1 + p^2 y_2 + \cdots\right)}{dt^2}\right.$$

$$-\lambda^2\left(y_0 + py_1 + p^2 y_2 + \cdots\right)$$

$$\left. -\frac{\lambda^4\left(y_0 + py_1 + p^2 y_2 + \cdots\right)^3}{6}\right] = 0$$

(E4)

Comparing the coefficients of like powers of p in Equation (E4) we get

$$p^0 : \frac{d^2 y_0}{dt^2} - \lambda^2 y_0 = 0$$

(E5)

$$p^1 : \frac{d^2 y_1}{dt^2} - \lambda^2 y_1 - \frac{\lambda^4 y_0^3}{6} = 0$$

(E6)

The initial approximations are as follows

$$y_0(0) = 0, \ y_0(1) = 1,$$

(E7)

$$y_i(0) = 0, \ y_i(1) = 0, \ i = 1, 2, 3 \cdots$$

(E8)

Solving the Equation (E5) and the Equation (E6) and using the boundary conditions Equation (E7) and the Equation (E8) we obtain the following results:

$$y_0 = \left(\frac{\sinh(\lambda t)}{\sinh(\lambda)}\right)$$

(E9)

$$y_1 = \left(\frac{\lambda^3}{48\sinh^3(\lambda)}\right)$$

$$\cdot\left[\left(\frac{\sinh(\lambda t)}{\sinh(\lambda)}\right)\left(3\cosh(\lambda) - \frac{\sinh(3\lambda)}{4}\right)\right.$$

$$\left. +\left(\frac{\sinh(3\lambda t)}{4\lambda} - 3t\cosh(\lambda t)\right)\right]$$

(E10)

According to the HPM, we can conclude that

$$y = \lim_{p \to 1} y(t) = y_0 + y_1$$

(E11)

After putting the Equation (E9) and the Equation (E10) into an Equation (E11) we obtain the solution in the text.

Appendix G: Matlab Program Is to Find the Numerical Solution of the Non Linear Differential Equations (15) and (16)

```
function pdex4 m = 0;
x = linspace(0,1);
t=linspace(0,10000);
sol= pdepe(m,@pdex4pde,@pdex4ic,@pdex4bc,x,t);
u1 = sol(:,:,1);
figure plot(x,u1(end,:))
title('u1(x,t)')
xlabel('Distance x')
ylabel('u1(x,2)')
%-----------------------------------------------------------------
function [c,f,s] = pdex4pde(x,t,u,DuDx)
c = 1;
f = DuDx;
lamda=2.8;
F =-lamda*(sinh(lamda*u))
s = F;
%-----------------------------------------------------------------
function u0 = pdex4ic(x); %create a initial conditions u0 = 1;
%-----------------------------------------------------------------
function[pl,ql,pr,qr]=pdex4bc(xl,ul,xr,ur,t) %create a boundary conditions pl
= ul;
ql = 0;
pr = ur-1;
qr = 0;
```

Appendix H: Solution of Catalytic Reactions in a Flat Particle Using HPM

In this Appendix, we indicate how the Equation (22) is derived. When $\dfrac{\gamma\beta(1-y)}{1+\beta(1-y)}$ is small, Equation (20) is reduces to

$$\frac{d^2 y}{dt^2} - \left(\lambda + \frac{\lambda\gamma\beta}{(1+\beta)}\right)y + \left(\frac{\lambda\gamma\beta y^2}{(1+\beta)^2}\right) = 0$$

(H1)

We construct the Homotopy for the Equation (H1) is as follows:

$$(1-p)\left[\frac{d^2 y}{dt^2} - \lambda\left(1+\frac{\gamma\beta}{(1+\beta)}\right)y\right]$$
$$+ p\left[\frac{d^2 y}{dt^2} - \lambda\left(1+\frac{\gamma\beta}{(1+\beta)}\right)y + \frac{\lambda\gamma\beta y^2}{(1+\beta)^2}\right] = 0$$

(H2)

The analytical solution of Equation (20) with Equation (21) is

$$y = y_0 + py_1 + p^2 y_2 + \cdots$$

(H3)

Substituting the Equation (E3) into an Equation (E2) we get

$$(1-p)\left[\frac{d^2\left(y_0 + py_1 + p^2 y_2 + \cdots\right)}{dt^2}\right.$$
$$\left. - \lambda\left(1+\frac{\gamma\beta}{(1+\beta)}\right)\left(y_0 + py_1 + p^2 y_2 + \cdots\right)\right]$$
$$+ p\left[\frac{d^2\left(y_0 + py_1 + p^2 y_2 + \cdots\right)}{dt^2}\right.$$
$$- \lambda\left(1+\frac{\gamma\beta}{(1+\beta)}\right)\left(y_0 + py_1 + p^2 y_2 + \cdots\right)$$
$$\left. + \frac{\lambda\gamma\beta\left(y_0 + py_1 + p^2 y_2 + \cdots\right)^2}{(1+\beta)^2}\right] = 0$$

(H4)

Comparing the coefficients of like powers of p in Equation (H4) we get

$$p^0 : \frac{d^2 y_0}{dt^2} - \lambda\left(1+\frac{\gamma\beta}{(1+\beta)}\right)y_0 = 0$$

(H5)

$$p^1 : \frac{d^2 y_1}{dt^2} - \lambda\left(1+\frac{\gamma\beta}{(1+\beta)}\right)y_1 + \left(\frac{\lambda\gamma\beta y_0^2}{(1+\beta)^2}\right) = 0$$

(H6)

The initial approximations are as follows

$$y_0'(0) = 0, \ y_0(1) = 1,$$
(H7)

$$i = 1, 2, 3 \cdots \ y_i'(0) = 0, \ y_i(1) = 0,$$
(H8)

Solving the Equation (H5) and the Equation (H6) and using the boundary conditions Equation (H7) and the Equation (H8) we obtain the following result:

$$y_0 = \left(\frac{\cosh(kt)}{\cosh(k)} \right)$$
(H9)

$$y_1 = \left(\frac{\lambda \, \beta \, \gamma \left(\cosh(2k) - 3 \right)}{6k^2 (1+\beta)^2 \cosh^2(k)} \right) \left(\frac{\cosh(kt)}{\cosh(k)} \right)$$

$$+ \frac{\lambda \gamma \beta \left[3 - \cosh(2kt) \right]}{6(1+\beta)^2 k^2 \cosh^2(k)}$$
(H10)

where k is defined in the text Equation (23).

According to the HPM, we can conclude that

$$y = \lim_{p \to 1} y(t) = y_0 + y_1$$
(H11)

After putting the Equation (H9) and the Equation (H10) into an Equation (H11) we obtain the solution in the text.

Appendix I: Matlab Program Is to Find the Numerical Solution of the Non Linear Differential Equations (20) and (21)

```
function pdex4 m = 0;
x =linspace(0,1);
t=linspace(0,10000);
sol= pdepe(m,@pdex4pde,@pdex4ic,@pdex4bc,x,t);
u1 = sol(:,:,1);
figure plot(x,u1(end,J)
title('u1(x,t)')
xlabel('Distance x')
ylabel('u1(x,2)')
%------------------------------------------------------------
function [c,f,s] = pdex4pde(x,t,u,DuDx)
```

```
c = 1;
f = DuDx;
lamda=4;
beta=0.1;
gamma=5;
F=-lamda*u*exp(beta*gamma*(1-u)/(1+beta*(1-u)));
s = F;
%-----------------------------------------------------------------
function u0 = pdex4ic(x); %create a initial conditions u0 = 1;
% -----------------------------------------------------------------
function[pl,ql,pr,qr]=pdex4bc(xl,ul,xr,ur,t) %create a boundary conditions pl
= 0;
ql = 1;
pr = ur(1)-1;
qr = 0;
```

Appendix J: Matlab Program Is to Find the Numerical Solution of the Non Linear Differential Equations (20) and (21)

```
function pdex4 m = 0;
x =linspace(0,1);
t=linspace(0,10000);
sol= pdepe(m,@pdex4pde,@pdex4ic,@pdex4bc,x,t);
u1 = sol(:,:,1);
figure plot(x,u1(end,:))
title('u1(x,t)')
xlabel('Distance x')
ylabel('u1(x,2)')
%-----------------------------------------------------------------
function [c,f,s] = pdex4pde(x,t,u,DuDx)
c = 1;
f = DuDx;
lamda=1;
beta=0.15;
```

```
gamma=10;
F=-lamda*u*exp(beta*gamma*(1-u)/(1+beta*(1-u)));
s = F;
%------------------------------------------------------------------
function u0 = pdex4ic(x); %create a initial conditions u0 = 1;
%------------------------------------------------------------------
function[pl,ql,pr,qr]=pdex4bc(xl,ul,xr,ur,t) %create a boundary conditions pl
= 0;
ql = 1;
pr = ur(1)-1;
qr = 0;
```

Appendix K: Matlab Program Is to Find the Numerical Solution of the Non Linear Differential Equations (20) and (21)

```
function pdex4 m = 0;
x =linspace(0,1);
t=linspace(0,10000);
sol= pdepe(m,@pdex4pde,@pdex4ic,@pdex4bc,x,t);
u1 = sol(:,:,1);
figure plot(x,u1(end,:))
title('u1(x,t)')
xlabel('Distance x')
ylabel('u1(x,2)')
%------------------------------------------------------------------
function [c,f,s] = pdex4pde(x,t,u,DuDx)
c = 1;
f = DuDx;
lamda=1;
beta=0.1;
gamma=15;
F=-lamda*u*exp(beta*gamma*(1-u)/(1+beta*(1-u)));
s = F;
%------------------------------------------------------------------
```

function u0 = pdex4ic(x); %create a initial conditions u0 = 1;

%---

function[pl,ql,pr,qr]=pdex4bc(xl,ul,xr,ur,t) %create a boundary conditions pl = 0;

ql = 1;

pr = ur(1)-1;

qr = 0;

Appendix: L Nomenclature

Symbol	Meaning
t	Dimensionless distance down the reactor
y	Dimensionless concentration in the reactor
λ	Dimensionless parameter
α	Dimensionless parameter
β	Dimensionless parameter
γ	Dimensionless parameter

Chapter 3

COMPUTATIONAL METHODS IN THE THEORY OF SYNTHESIS OF RADIO AND ACOUSTIC RADIATING SYSTEMS

Petro Savenko

Department of Numerical Methods of the Pidstryhach Institute for Applied Problems of Mechanics and Mathematics, National Academy of Sciences of Ukraine, Lviv, Ukraine

ABSTRACT

A brief review of the works of the author and his co-authors on the application of nonlinear analysis, numerical and analytical methods for solving the nonlinear inverse problems (synthesis problems) for optimizing the different types of radiating systems, is presented in the paper. The synthesis problems are formulated in variational statements and further they are reduced to research and numerical solution of nonlinear integral equations of Hammerstein type. The existence theorems are proof, the investigation methods of nonuniqueness problem of solutions and numerical algorithms of finding the optimal solutions are proved.

INTRODUCTION

In many practical applications at the optimal design of various types of radio and acoustic radiating systems the requirements are only to the energy characteristics of the directivity of the radiated field (amplitude directivity pattern (DP) or DP by the power). Therefore there is a need to approximate real finite functions by modules of onedimensional or two-dimensional and discrete Fourier transform dependent on the real physical parameters. At the same time the absence of requirements to phase characteristics of field is used to improve the quality of approximation of synthesized DP to given.

Later on the variational formulations of different types of inverse problems in mean-square approach, which in further are reduced to investigation and numerical solution of one-dimensional or two-dimensional nonlinear integral equations of the Hammerstein type with separate module and argument of desired complex-valued function, are considered. Nonuniqueness and branching (or bifurcation) of solutions dependent on the change of the physical parameters characterizing the radiating system are characteristic features of such equations. Problems on finding the set of branching points (bifurcation) are not investigated nonlinear one-parameter or two-parameter spectral problems.

The existence of connected components of the spectrum, which in the case of real parameters are of spectral lines, is essential difference between the two-dimensional and one-dimensional spectral problems. The problem on finding the spectral lines is reduced to numerical solution of the Cauchy problem for an ordinary differential equation of the first order.

The degenerate of kernels in linear operators of the Hammerstein type equations is feature of the synthesis problems of antenna arrays. It allows to reduce nonlinear two-parameter spectral problems on finding the set of branching points of solutions to the corresponding systems of linear algebraic equations with nonlinear occurrence of the spectral parameters in the coefficients of system.

In the basis of construction of numerical algorithms for finding the optimal solutions are taken such princeples: localization of existing solutions dependent on the value of the physical parameters of the problem by means the use of numerical methods of solving the non-linear one-parameter and two-parametric spectral problems, and methods of the branching theory of solutions—construction and justification of the convergence of iterative processes for numerical finding the various types of existing solutions of basic equations (equations of Hammerstein type)—analysis of the effectiveness of found solutions.

FORMULATION OF PROBLEMS. BASIC EQUATIONS OF SYNTHESIS

In general case, the analysis problems (direct problems) of radio (or acoustic) radiating systems are reduced to solution the corresponding boundary problems of electrodynamics (acoustic) at a given excitation sources of fields [1-4] on the basis of Maxwell's equations (wave equation). The directivity pattern f is one of the basic characteristics of the emitted field on large distances from the radiating system. It describes the properties of the field in space dependent on the angular coordinates of a spherical coordinate system. In general case, the DP f is a vector of complex-valued function which has the form [5,6]

$$f = f_\vartheta i_\vartheta + f_\varphi i_\varphi \equiv A_\vartheta U i_\vartheta + A_\varphi U i_\varphi.$$ (1)

Here $A = \{A_\vartheta, A_\varphi\}$ is linear operator acting from some functional Hilbert space H_υ (the space of square integrable functions in the domain V, describing the distribution of extraneous fields (currents) U in volume v) into the space of complex-valued continuous functions $C(\Omega)$ defined in some domain $\Omega \subseteq \mathsf{U}^2$ (or $\Omega \subseteq \mathsf{\Pi}^1$) [7]. The form and properties of operators A_ϑ, A_φ are defined by type and geometry of the radiating system. The set (domain) of values of the operator $A = \{A_\vartheta, A_\varphi\}$ is called [8,9] set or class of realized directivity patterns. This means that for any DP f from this class there exists such function of distribution of the currents (fields) $U \in H_\upsilon$ that realizes this DP, i.e. $AU = f$.

In the simplest form the inverse problem (the synthesis problem) according to the prescribed amplitude DP can be formulated as the problem on finding the solutions of nonlinear operator equation of the first kind

$$|AU| = F,$$ (2)

where F is a given amplitude DP. In staged thus the synthesis problem all three correct conditions of problems by Hadamard [10-12]: existence of the solution, uniqueness of the solution, continuous dependence of the solution of the input data, can be violated simultaneously. Violation of condition (1) in the first place is connected with the fact that the given DP F can not belong to the class realized, that is to the domain of values of the operator A. In other words, such DP can not be obtained at any distribution of field in the aperture of the radiating system belonging to the space H_υ. Trying to recreate the DP F just leads to effect of superdirectivity [5]. The system becomes resonance and critical to change of parameters.

Condition (2) is violated due to the nonlinearity of the problem.

Therefore, the variational formulations of problems, which in addition to the requirements of the basic characteristics of DP also contain requirements to the distribution function of the currents (fields) in the aperture of the radiating system, are considered. At that is required not complete coincidence obtained DP $|f|$ with given F, but only the best (in the sense of the selected criterion) approximation to it.

An important feature of the variational formulation of synthesis problems is the fact that in the optimization criterion can introduce functionals describing certain other requirements to amplitude-phase distribution (APD) of outside excitation sources. Their mean-square deviation, as a rule, will be used as the criterion of proximity of amplitudes of the given and synthesized DP.

The Case of Linear Polarization of Extraneous Field

First we consider the scalar case of problems when extraneous fields (currents) in the radiating system is linearly polarized [7,13], and created by their DP (1) has only one component. Let the operator A acts from some Hilbert functional space $H_U = L_2(V)$ into the complex space of continuous and square integrable functions in domain $\Omega \subseteq \sqcup^2$ (or $\Omega \subseteq \sqcup^1$) $C_{(\Omega)}^{(2)}$.

In space H_U we introduce the scalar product and norm

$$(U_1, U_2)_{H_U} = \int_V U_1(P) \cdot \overline{U_2(P)} dV$$,

$$\|U\|_{H_U} = \left\{ \int_V |U(P)|^2 dV \right\}^{1/2}$$ (3)

where $P = (x, y, z)$ is a point of integration.

Along with the Chebyshev norm

$$\|f\|_{C(\Omega)} = \sup_{Q \in \Omega} |f(Q)|$$,   (4)

in the space we introduce scalar product and the generated by it norm and metric $C_{(\Omega)}^{(2)}$ as follows:

$$(f, g)_{C_{(\Omega)}^{(2)}} = \int_\Omega f(Q) \overline{g(Q)} dQ, \quad \|f\|_{C_{(\Omega)}^{(2)}} = (f, f)_{C_{(\Omega)}^{(2)}}^{1/2}$$,

$$P_{C_{(\Omega)}^{(2)}} (f_1, f_2) = \|f_1 - f_2\|_{C_{(\Omega)}^{(2)}}$$.   (5)

Note, space $C_{(\Omega)}^{(2)}$ is a Banach space relatively uniform norm (4) and it is incomplete space concerning norm defined according to (5) [14].

We consider also that the operator A has the izometric property or it is completely continuous. Let the given amplitude DP F is real positive (nonnegative) continuous function which different from nonzero in some limited closed domain $\bar{G} \subset \Omega$ and identically equal zero on complement Ω $\bar{G}$. Let A is isometric operator, that is for any $U \in H_U$ and $f = AU \in H_f$ equality is satisfied

$$\|f\|^2_{C^{(2)}_{(\Omega)}} = (AU, AU)_{C^{(2)}_{(\Omega)}} = (U,U)_{H_U} = \|U\|^2_{H_U} . \tag{6}$$

In [15-17] the synthesis problem of given amplitude DP F is formulated (is investigated) as a minimization problem in the Hilbert space $H_U = L_2(V)$ of the functional

$$\sigma_F(U) = \big\| F - |AU| \big\|^2_{C^{(2)}_{(\Omega)}} \tag{7}$$

characterizing the value of mean-square deviation of modules of given and synthesized DP in domain Ω. For the formulated problem occurs [18,19]

Theorem 2.1. Let the linear operator

$A: L_2(V) \rightarrow C^{(2)}_{(\Omega)}$ is isometric relatively mean-square metric and it is continuous operator from $L_2(V)$ into

$C^{(2)}_{(\bar G)}$ concerning uniform metric of the space $C^{(2)}_{(\Omega)}$, and the given amplitude DP $F(s)$ is a real positive (nonnegative) continuous function in the domain $\bar G$.

Then at least one point of absolute minimum of the functional (7) exists in the space $U = A^* (F \cdot e^{i \arg AU})$ and a subsequence which converges weakly to one of points of absolute minimum can be selected from any minimizing sequence.

On the base of the necessary condition for an extremum of the functional (zero equality of its Hato differential [20]) $D\sigma_F(U, w) = \sigma'_F(U)(w) \equiv 0$, we obtain the equation with respect to the optimal distribution of excitation sources

$$U = A^* \left(F \cdot e^{i \arg AU} \right), \tag{8}$$

where A^* is the conjugate with A operator.

Let the set of zeros $N(A)$ of the operator A consists of only one zero element 0. Then acting on both parts of (8) by operator A, we obtain the equivalent to (8) equation in space $C(\bar G)$:

$$f = AA^* \left(F \cdot e^{i \arg f} \right). \tag{9}$$

By solutions f_* of this equation the optimal distribution of excitation sources in radiating system are defined by the formula

$$U_* = A^* \left(F \cdot e^{i \arg f_*} \right). \tag{10}$$

From Theorem 2.1 and the properties of functional $\sigma_F(U)$ follows Corollary

2.1. Since functional $\sigma_F(U)$ is differentiable in H_U by Hato, it is growing [21] and according to Theorem 2.1 has at least one point of absolute minimum, then (8) in the space H_U and (9) in the space $C(\bar{G})$ have at least one solution.

Lemma 2.1. Between solutions of (8) and (9) there exists bijection, that is if U_* is the solution of (8), then $f_* = AU_*$ is the solution of (9). On the contrary, if f_* is the solution of (9), then the corresponding solution of (8) is defined by (10).

The possibility of investigation of solutions of synthesis problems, using (8) or (9) follows from Lemma. Note, Equation (9) is simpler as (8), since the latter contains the operator exponent.

Note, solutions of synthesis problems according to the prescribed amplitude DP are determined with precision to value $e^{i\beta}$ (β is arbitrary constant), since $\left| Ue^{i\beta} \right| = |U|$. So if there exists the solution of Equations (8) (or (9)), then there is also generated by its family of solutions in which one solution different from another by phase constant. For the uniqueness of desired solutions additional conditions impose on the functions $\arg U$ or $\arg f$.

In the case completely continuous operator describing DP of radiating system the smoothing Tikhonov type functional [11]

$$\sigma_{F_\alpha}(1) = \left\| F - |f| \right\|^2_{C^{(2)}_{(G)}} + \alpha \|U\|^2_{H_U}$$
$$\equiv \left\| F - |AU| \right\|^2_{C^{(2)}_{(G)}} + \alpha \|U\|^2_{H_U} \tag{11}$$

which includes requirements as to the mean-square deviation of DP, so to norm of current, is used [18,22] for the synthesis of various types of antennas. The parameter α can be viewed as regularization parameter [11,23] or as a weighing coefficient, by means of which can control ratio between the first and second summand of functional.

Theorem 2.2. Let the linear operator A acts from the complex Hilbert space $H_U = L_2(V)$ into the complex space of continuous functions $C^{(2)}_{(G)}$ and it is completely continuous, and given DP is real positive (nonnegative) and continuous function in $\hat{G} \in \mathbb{R}^1$ (or in $\bar{G} \in \mathbb{R}^2$)).

Then at least one point of absolute minimum of functional $\sigma_{F_\alpha}(U)$ exists in $H_U = L_2(V)$ and a subsequence which converges weakly to one of the points of absolute minimum can be selected from any minimizing sequence.

Differentiating functional $\sigma_{F_\alpha}(U)$ by Hato and performing appropriate transformations, we obtain the equation [22,24]

$$\alpha U = -A^* A U + A^* \left(F \exp\left(i \arg\left(A U \right)\right)\right) \tag{12}$$

in the space H_U.

Equation with respect to synthesized DP based on equality $f = AU$ and (12) has the form

$$\alpha f = -A A^* f + A A^* \left(F \exp\left(i \arg f\right)\right) \tag{13}$$

Lemma similar to Lemma 2.1 is valid for (12) and (13).

From Theorem 2.2 and properties of functional $\sigma_{F_\alpha}(U)$ follows [7,18]

Corollary 2.2. Since differentiable in H_U by Hato functional $\sigma_{F_\alpha}(U)$ is growing and according to Theorem 2.2 has at least one point of absolute minimum, then (12) in space H_U and (13) in the space $C(\bar{G})$ have at least one by one solution.

The Case of Arbitrary Polarization of Excitation Fields

Consider the more general case when the excitation fields (or currents) in the radiating system and generated by it DP have vector character [5,6]. In this case, we set that the operator A is completely continuous and it acts from $H_U = L_2(V) \otimes L_2(V) \otimes L_2(V)$ complex space of square integrable in the domain V vector-valued functions, into the complex space of continuous functions on the compact $\bar{G} \subset \mathbb{R}^2$ vector-valued functions

$C_{(\bar{G})}^{(2)} = C_{(\bar{G})}^{(2)} \otimes C_{(\bar{G})}^{(2)}$ equipped by scalar product. We introduce the scalar product and generated by it norm in H_U:

$$(U_1, U_2) = \left(U_x^{(1)}, U_x^{(2)}\right) + \left(U_y^{(1)}, U_y^{(2)}\right) + \left(U_z^{(1)}, U_z^{(2)}\right)$$

$$\equiv \int_V \left\{ U_x^{(1)}(x,y,z)\overline{U_x^{(2)}}(x,y,z) + U_y^{(1)}(x,y,z)\overline{U_y^{(2)}}(x,y,z) \right.$$

$$\left. + U_z^{(1)}(x,y,z)\overline{U_z^{(2)}}(x,y,z) \right\} dxdydz, \tag{14}$$

$$\|U\|_{H_U} = (U,U)^{1/2}$$

$$= \left(\left(U_x, U_x\right) + \left(U_y, U_y\right) + \left(U_z, U_z\right)\right)^{1/2} \tag{15}$$

We define module of vector U as following:

$$|U| = \left(U_x \cdot \bar{U}_x + U_y \cdot \bar{U}_y + U_z \cdot \bar{U}_z\right)^{1/2}.$$

In the space $C^{(2)}_{(\bar{G})} = C^{(2)}_{(\bar{G})} \otimes C^{(2)}_{(\bar{G})}$ along with the Chebyshev norm

$$\|f\|_{C(\bar{G})} = \max_{(\vartheta,\varphi) \in G} |f(\vartheta,\varphi)|, \tag{16}$$

where

$$|f(\vartheta,\varphi)| = \left(f_\vartheta(\vartheta,\varphi) \cdot \overline{f_\vartheta(\vartheta,\varphi)} + f_\varphi(\vartheta,\varphi) \cdot \overline{f_\varphi(\vartheta,\varphi)}\right)^{1/2}$$ we introduce the scalar product and generated by it the mean-square norm and metric

$$(f_1, f_2)_{C^{(2)}_{(\bar{G})}} = \left(f_\vartheta^{(1)}, f_\vartheta^{(2)}\right)_{C^{(2)}_{(\bar{G})}} + \left(f_\varphi^{(1)}, f_\varphi^{(2)}\right)_{C^{(2)}_{(\bar{G})}}$$

$$\equiv \iint_G \left\{ f_\vartheta^{(1)}(\vartheta,\varphi) \cdot \overline{f_\vartheta^{(2)}}(\vartheta,\varphi) + f_\varphi^{(1)}(\vartheta,\varphi) \cdot \overline{f_\varphi^{(2)}}(\vartheta,\varphi) \right\}$$

$$\cdot \sin\vartheta\, d\vartheta\, d\varphi, \tag{17}$$

$$\|f\|_{C^{(2)}_{(\bar{G})}} = (f, f)^{1/2}_{C^{(2)}_{(\bar{G})}}. \tag{18}$$

If DP of radiating system has two components f_ϑ, f_φ, i.e. it is described by (1), then as the optimization criterions are used the following functionals [7]:

$$\sigma_{F_\alpha}(U) = \left\| F - |f| \right\|^2_{C^{(2)}_{(\bar{G})}} + \alpha \|U\|^2_{H_U}$$

$$\equiv \left\| F - |AU| \right\|^2_{C^{(2)}_{(\bar{G})}} + \alpha \|U\|^2_{H_U}, \tag{19}$$

$$\tilde{\sigma}_{F_\alpha}(U) = \left\| F_\vartheta - |f_\vartheta| \right\|^2_{C^{(2)}_{(\bar{G})}} + \left\| F_\varphi - |f_\varphi| \right\|^2_{C^{(2)}_{(\bar{G})}} + \alpha \|U\|^2_{H_U}$$

$$\equiv \left\| F_\vartheta - |A_\vartheta U| \right\|^2_{C^{(2)}_{(\bar{G})}} + \left\| F_\varphi - |A_\varphi U| \right\|^2_{C^{(2)}_{(\bar{G})}} + \alpha \|U\|^2_{H_U}. \tag{20}$$

In the functional (20), F_ϑ, F_φ are the given amplitude of components of required DP. At that

$$F = \left(F_\vartheta^2 + F_\varphi^2\right)^{1/2}, \quad \|F\|_{C(\bar{G})} = 1$$ and functions F_ϑ, F_φ

can be given with account of existing requirements to polarization characteristics of emitted field.

If in the synthesis problem functional (19) is used as the optimization criterion, the problem on finding the minimum points is reduced to finding the solutions of the equation

$$U = -\alpha^{-1} A^* AU + \alpha^{-1} A^* \left(F\, AU / |AU| \right) \tag{21}$$

in the space H_I. Equivalent to (21) equation with respect to vector DP f in space $C_{(\bar{G})}^{(2)}$ has the form

$$f = -\alpha^{-1}AA^*f + \alpha^{-1}AA^*\left(F\,f/|f|\right)_.$$

(22)

In this case the following theorem is valid [7].

Theorem 2.3. Let linear completely continuous operator A acts from the complex Hilbert space

$$H_U = L_2(V)\otimes L_2(V)\otimes L_2(V)$$ into the complex space of continuous functions $C_{(\bar{G})}^{(2)} = C_{(\bar{G})}^{(2)} \otimes C_{(\bar{G})}^{(2)}$ equipped by the scalar product, and given DP is a real positive continuous function on the compact $\bar{G}$.

Then in H_U there exists at least one point of absolute minimum of functional $\sigma_{F_\alpha}(U)$ and a subsequence which converges weakly to one of the points of absolute minimum can select from any minimizing sequence.

For the functional (20) is true Theorem 2.4. At conditions of Theorem 2.3 functional $\tilde{\sigma}_{F_\alpha}(U)$ in the space H_U has at least one point of absolute minimum and subsequence which converges weakly to one of points of absolute minimum can be selected from any minimizing sequence.

For minimizing of the functional $\tilde{\sigma}_{F_\alpha}(U)$ in the space H_U we obtain equation [7]

$$U = \alpha^{-1}A^*\left\{\left(F_g - |A_gU|\right)e^{i\arg A_gU},\left(F_\varphi - |A_\varphi U|\right)e^{i\arg A_\varphi U}\right\}_.$$

(23)

Equivalent to (23) equation with respect to synthesized DP f in the space $C_{(\bar{G})}^{(2)}$ has the form

$$f = \frac{1}{\alpha}AA^*\left\{\left(F_g - |f_g|\right)e^{i\arg f_g},\left(F_\varphi - |f_\varphi|\right)e^{i\arg f_\varphi}\right\}_.$$

(24)

The existence of solutions of (23) in space H_U and (24) in the space $C_{(\bar{G})}^{(2)}$, respectively, follows from necessary condition of functional minimum $\tilde{\sigma}_{F_\alpha}(U)$.

If necessary the weight function [15,17] can introduce in the functionals σ_F and σ_{F_α} by means entering appropriate scalar products and affect on quality of the approximation of synthesized and given DP in a certain range of angles.

Simultaneous Optimization of the Geometry of Aperture and Excitation Fields

The synthesis problems with optimization of geometry of radiating system are more complicated class of problems. These problems need to find a configuration of the radiating system and amplitude-phase distribution of excitation fields (currents) in it [25,26]. Moreover, the operator A depends on two functions: function Γ describing the geometry of the system, and amplitude-phase distribution function of excitation sources U, i.e.

$$f = A(\Gamma)U.$$

$$(25)$$

In addition, the function U has, as rule, vector character, and the operator A is a nonlinear concerning the function Γ. Later on we shall consider the synthesis problem of a flat aperture, in which in addition to amplitude-phase distribution (APD) desired is too the function that describes the boundary of aperture. The basis of the formulation of such problems can be put the functionals (7), (11) and (20), expanding their by corresponding requirements to geometry of radiating system.

Synthesis Problem of Discrete Radiating Systems Antenna Arrays

In many radio engineering systems antenna arrays (AR) have gained widespread use. Antenna array is [4,6,27] antenna, which consists of N identical (or differenttype) radiators placing corresponding way in space and they collate by common system of power and control. In [28-37] investigations of nonlinear synthesis problems and planar antenna arrays according to the prescribed amplitude DP are presented.

To describe the electromagnetic characteristics of antenna arrays are used different by precision mathematical models [38-42]. In the base of construction of mathematical models is imposed [40,42] that the excitation of each radiator is characterized by a unique complex number I_n-complex amplitude of excitation. It's the physical meaning depends on the type of radiating system. On the base on the linearity of the Maxwell's equations the complex excitation amplitudes enter linearly in the expression for DP of array, that is

$$f(\vartheta,\varphi) = \sum_{n=1}^{N} I_n f_n(\vartheta,\varphi) e^{ik(x_n \sin\vartheta\cos\varphi + y_n \sin\vartheta\sin\varphi + z_n \cos\vartheta)}$$

$$,\qquad (26)$$

where $f_n(\vartheta,\varphi) = f_\vartheta^{(n)}(\vartheta,\varphi)i_\vartheta + f_\varphi^{(n)}(\vartheta,\varphi)i_\varphi$ is vector DP n-th radiator. Vector $I = \{I_1, I_2, \cdots, I_N\}$ is called vector excitation of array or vector of amplitude-phase distribution of currents in the array.

In general, the construction of high-accuracy mathematical models of array is reduced to solving the corresponding exterior boundary problem of high-frequency electrodynamics for system of the Maxwell's equations in multiply-connected domain [1,39-41]. In the particular case, where the elements of the array are ideally leading talamy accounting of the mutual influence is based on the method of induced electromotive forces (IEF) and it is reduced to solution the corresponding system of linear integral equations [42]

$$BI = U,$$
(27)

where B is matrix-integral linear operator; I is complexvalued vector distribution function of surface currents on radiators; $U \in H_U$ is vector-valued function describing the outside fields (voltage) which is necessary to create in the system of power of array. Allocating in the space H_I compact class of solutions (where (27) is correct), solution of (27) is written as

$$I = \tilde{B}^{-1}U.$$
(28)

Here it is assumed that the corresponding regularized system[1] $\tilde{B}I = U$ exists for the system of (27). Then on the basis of (28) formula for DP of array takes the form

$$f = A\tilde{B}^{-1}U.$$
(29)

This relation allows to formulate the synthesis problem of antenna array according to the prescribed amplitude DP with account of the mutual influence of elements as the problem on finding the vector U minimizing the functional

$$\sigma_{F_\alpha}(U) = \left\| F - \left| A\tilde{B}^{-1}U \right| \right\|^2_{C^{(2)}_{(G)}} + \alpha \|U\|^2_{H_U}$$
(30)

in space H_U.

At need to take into account the component-wise approximation of modules of given and synthesized DP's, the functional [43-45]

$$\tilde{\sigma}_{F_\alpha}(U) = \left\| F_g - |f_g| \right\|^2_{C^{(2)}_{(G)}} + \left\| F_\varphi - |f_\varphi| \right\|^2_{C^{(2)}_{(G)}} + \alpha \|U\|^2_{H_U}$$

$$\equiv \left\| F_g - \left| A_g\tilde{B}^{-1}U \right| \right\|^2_{C^{(2)}_{(G)}} + \left\| F_\varphi - \left| A_\varphi\tilde{B}^{-1}U \right| \right\|^2_{C^{(2)}_{(G)}}$$

$$+ \alpha \|U\|^2_{H_U}$$
(31)

analogously to (20) can be used as optimizing criterion.

By the desired solutions of this problem the optimal vector of extraneous voltages on inputs of radiators is determined on the basis of (27). Thus, the basic requirements for synthesis problems of different types of radiating systems according to the prescribed amplitude DP are formulated. Note that recorded functionals is not convex [21], and therefore may have nonunique extreme point. In further the above statements of problems allow to obtain relatively simple nonlinear integral or matrix equations for the study and solution of which can be applied numerical methods of nonlinear analysis methods.

Integral equations method [40,42] is used widely in such classes of problems. The method of synthesis of antenna arrays from cylindrical dipoles with account of mutual influence is proposed in [43-45]. Analysis of problem of nonuniqueness solutions is studied there by means computational experiments.

Nonlinear Synthesis Problem of Radiating Systems with Use of Energy Criterion

In spite of the fact that from the amplitude DP $|f|$ is easy to obtain the DP by power $N = |f|^2$ and vice versa, in the mathematical aspect the synthesis problems of given amplitude DP F_0 and given energy DP $N_0 = F_0^2$ are different tasks. For example, if $|f_*|$ is the optimal solution of some variational synthesis problem of amplitude DP, then $|f_*|^2$ will not be the optimal solution of a similar problem for the given DP F_0^2. On this basis, in [46-49] on the operator level are considered statements of synthesis problems with use of two types of stabilizing functionals, in which the vector character of the electromagnetic fields takes into account.

Consider the synthesis problem of given energy DP $N_0(\vartheta, \varphi)$. Taking into account the expression for DP

$N = |f|^2 = |f_\vartheta|^2 + |f_\varphi|^2 \equiv |A_\vartheta U|^2 + |A_\varphi U|^2$ in the simplest form this problem can be formulated as a problem on finding the solutions of nonlinear operator equation of the first kind

$$|AU|^2 \equiv |A_\vartheta U|^2 + |A_\varphi U|^2 = N_0,$$
(32)

where N_0 is a real positive continuous function in

$\bar{G} \in \sqcup^2$ (at that $\max_{(\vartheta, \varphi) \in \bar{G}} N_0(\vartheta, \varphi) = 1$) which can not belong to the domain of values of the operator $|AU|^2$. It is known [11] that (32) is severely ill-posed problem. In this connection, we consider the problem on best meansquare approximation of the real positive continuous (in the domain $\overline{G}$) function $N_0(\vartheta,\varphi)$ by function $|f(\vartheta,\varphi)|^2$

$$(f(\vartheta,\varphi) = AI \in R(A), \quad U = \{U_x, U_y, U_z\} \in H_U).$$

Formulate it as minimizing problem of functional [49]

$$\sigma_{N_\alpha}(U) = \left\| N_0 - |AU|^2 \right\|_{C_{(G)}^{(2)}}^2 + \alpha \|U\|_{H_U}^2$$

$$\equiv \left\| N_0 - |f|^2 \right\|_{C_{(G)}^{(2)}}^2 + \alpha \|U\|_{H_U}^2 \tag{33}$$

in the space H_U. In this functional the first summand characterizes mean-square deviation of given and synthesized DP by power. Second summand imposes restrictions on norm of currents in the radiating system. Real parameter $\alpha > 0$ we shall consider as a weighing multiplier. The existence of at least one point of minimum functional $\sigma_{N_\alpha}(U)$ in the space H_U states [7,49]

Theorem 2.5. Let the linear operator A acts from the space H_U into $C_{(\overline{G})}^{(2)}$ and it is completely continuous, $N_0(\vartheta,\varphi)$ is given nonnegative continuous the function in $\overline{G}$, at that $\max_{(\vartheta,\varphi)\in G} N_0(\vartheta,\varphi) = 1$.

Then in the space H_U there exists at least one point of absolute minimum of the functional $\sigma_{N_\alpha}(U)$ and subsequence that converges weakly to one of the points of absolute minimum can be selected from any minimizing sequence.

On the base of necessary condition of minimum functional is obtained the equation [7]

$$U = \frac{2}{\alpha} A^*(N_0 \cdot AU) - \frac{2}{\alpha} A^*(|AU|^2 \cdot AU) \tag{34}$$

with respect to optimal currents in the space H_U. This equation is a nonlinear operator equation having in the right part linear operator along with the Hammerstein type operator. If the set of zeros $N(A)$ consists of only the zero element 0, then acting on both parts of (34) by operator A, we obtain equation an equivalent to (34) with respect to synthesized DP in the space $C_{(\overline{G})}^{(2)}$

$$f = B(f) \equiv \frac{2}{\alpha} AA^*(N_0 \cdot f) - \frac{2}{\alpha} AA^*(|f|^2 \cdot f). \tag{35}$$

In [49] is shown that the functional $\sigma_{N_\alpha}(U)$ has m-property [50], that is the minimum point of the functional is interior point of some convex weakly closed set of the space H_U. On this basis from Theorem 2.5 follows [7]

Corollary 2.3. Since the functional $\sigma_{N_\alpha}(U)$ is differentiable in H_U by Hato, has at least one minimum point and m-property, then (34) in the space H_U and

(35) in the space $C^{(2)}_{(G)}$ have at least by one solution.

Lemma 2.2. At conditions of Theorem 2.5 and limited values of parameter α operator

$$B(f)=\frac{2}{\alpha}AA^{*}(N_{0}\cdot f)-\frac{2}{\alpha}AA^{*}\left(\left|f\right|^{2}\cdot f\right) \tag{36}$$

is compact in the space $C^{(2)}_{(\bar{G})}$.

Since for elements relatively compact set of normalized space the strong and weak convergence coincide [51] then with Theorem 2.5 and Lemma 2.2 follows Corollary 2.4. If $\{U_n\}$ is minimizing sequence of the functional $\sigma_{N_\alpha}(U)$ converging weakly to the minimum point U_*, then the sequence $\{f_n = AU_n\} \in C^{(2)}_{(\bar{G})}$ converges uniformly in $C^{(2)}_{(\bar{G})}$ to $f_* = AU_*$.

ABOUT BRANCHING OF SOLUTIONS OF THE BASIC EQUATIONS OF SYNTHESIS. PARTIAL CASES

Here on the example of scalar synthesis problems of linear radiator and radiating system with a flat aperture are presented the results of investigation of nonuniqueness problem of solutions corresponding to these tasks nonlinear integral equations of Hammerstein type depending on the change of the physical parameters.

The Case of a Linear Radiator

We put that the linear antenna is linear electric conductor of length $2a$, sizes of cross-section of which are much less than the wavelength. Due to these limitations the excitation currents in the antenna shall have only directaxis current [5]. Introduce the Cartesian and connected with spherical coordinate systems such that the origin of coordinates coincides with the middle of the antenna. We direct the axis OZ along the antenna. Then the current vector in a Cartesian coordinate system will have only z-component $U(z)$. We shall introduce the dimensionless coordinates $s = \sin\vartheta/\sin\gamma_0$, $z' = z/a$, and parameter

$$c = ka \sin \gamma_0,$$

(37)

connecting the electric size of antenna with angle $2\gamma_0$, outside of which given amplitude DP $F(s)$ identically zero. Then the formula for DP of linear antenna takes the form [7]

$$f(s) = AU \equiv \int_{-1}^{1} U(z') e^{icz's} dz'.$$

(38)

For DP of antenna the Parseval equality [8]

$$\int_{-\infty}^{\infty} |f(s)|^2 ds = (2\pi/c) \int_{-1}^{1} |U(z)|^2 dz$$

(39)

is valid, that is the operator A is isometric.

Taking into account that F is finite function with compact carrier $\bar{G} = [-1,1]$ and expressions for the operators A and A^*, we obtain the expanded form of (8):

$$U(z) = (c/2\pi) \int_{-1}^{1} F(s) \exp^{i\left\{ \arg \int_{-1}^{1} U(z') e^{icz's} dz' - czs \right\}} ds.$$

(40)

On the basis of (9), (38) we obtain the Hammerstein type equation concerning optimal DP

$$f(s) = Bf \equiv \int_{-1}^{1} F(s') K(s,s';c) e^{i \arg f(s')} ds'$$

(41)

in the space $C_{[-1,1]}^{(2)}$, where

$$K(s,s';c) = \sin c(s-s') / (\pi(s-s')).$$

(42)

The existence of at least one solution of (40) in the space $H_U = L_2[-1,1]$ and (41) in the space $C[-1,1]$ follows from Corollary 2.1.

We shall present three important properties of (41).

1) If $f(s)$ is the solution of (41), then complexconjugate function $\overline{f(s)}$ is the solution of (41) too.

2) If $f(s)$ is the solution of (41), then $e^{i\beta} f(s)$ is the solution of (41) too, where β is an arbitrary real constant.

3) For even functions $F(s)$ nonlinear operator B, which is in the right part of (41), transfers even phase DP $(\arg f(s))$ in even, and odd—in an odd.

That is the operator B is invariant with respect to the type of parity of function $\arg f(s)$. Due to this property the existence of fixed points of the operator B-solutions of (41) is possible in the classes of even and odd phase DP's.

In [16,52] is shown that (41) has two solutions in the class of real functions:

$$f_1(s,c) = \int_{-1}^{1} F(s') K(s,s';c) ds' \tag{43}$$

which is called the primary solution of the first type and

$$f_2(s,c) = \int_{-1}^{1} F(s') K(s,s';c) \operatorname{sgn}(s' - s_0) ds' \tag{44}$$

is the primary solution of the second type. Point $s = s_0(c)$ is determined from the condition $f_2(s,c) = 0$. For the even $F(s)$ $s_0 = 0$, that is the solution (44) is a real odd function (the corresponding to it amplitude DP is even function).

These solutions are effective only at small values of parameter c. With the growth of this parameter there exist branching points c_i at which more effective (in the sense value of functional σ_F) complex solutions branch-off from real solutions.

Consider according to [7,16] results of investigation of branchings of primary solution of the first type of solutions of (41). Using the procedure of decomplexification of the space $C[-1,1]$ [14] from (41) we move to the equivalent system

$$u(s) = \int_{-1}^{1} Q_1[s,s',c,u(s'),v(s')] ds'$$
$$\equiv \int_{-1}^{1} F(s') K(s,s';c) \frac{u(s')}{\sqrt{u^2(s') + v^2(s')}} ds',$$

$$v(s) = \int_{-1}^{1} Q_2[s,s',c,u(s'),v(s')] ds'$$
$$\equiv \int_{-1}^{1} F(s') K(s,s';c) \frac{v(s')}{\sqrt{u^2(s') + v^2(s')}} ds'. \tag{45}$$

On the base of the branching theory of solutions [53] the problem on finding such values of parameter $c = c_i$ (branching points) and all different from $f_1(s,c_i)$ solutions of system of (45) satisfying the conditions

$$\begin{cases} \max\limits_{s\in[-1,1]} |u(s,c_l) - f_1(s,c_l)| \to 0, \\ \max\limits_{s\in[-1,1]} |v(s,c_l)| \to 0, \end{cases} \quad \text{at } c \to c_l \tag{46}$$

are considered. Condition (46) means that it is necessary to find small continuous solutions

$$\omega(s) = v(s,c), \ w(s) = u(s,c) - f_1(s,c_l),$$ converging uniformly to zero at $c \to c_l$.

Putting in (45)

$$u(s,c) = f_1(s,c_l) + w(s), v(s) = 0 + \omega(s) \tag{47}$$

and expanding the integrand Q_1, Q_2 in the vicinity of the point $(c_l, f_1(s,c_l), 0)$ in the power series by w, ω and μ, and taking into account that the function $f_1(s,c)$ is its solution, we obtain system of integral equations of Lyapunov-Schmidt type [53] with respect to small solutions $w(s)$, $\omega(s)$:

$$w(s) = a(s,c_l)\mu$$

$$+ \sum_{m+n+p\geq 2}^{\infty} \mu^p \int_{-1}^{1} A_{mnp}(s,s') w^m(s') \omega^n(s') ds', \tag{48}$$

$$\omega(s) - \int_{-1}^{1} F(s') K(s,s';c) \frac{\omega(s')}{f_1(s',c)} ds'$$

$$= \sum_{m+n+p\geq 2}^{\infty} \mu^p \int_{-1}^{1} B_{mnp}(s,s') w^m(s') \omega^n(s') ds', \tag{49}$$

where $a(s,c_l) = \int_{-1}^{1} A_{001}(s,s') ds'$. On the base of the left part of (49) we obtain linear homogeneous integral equation

$$\omega(s) = T(c)\omega \equiv \int_{-1}^{1} F(s') K(s,s';c) \frac{\omega(s')}{f_1(s',c)} ds' \tag{50}$$

for finding the points of possible branching of solutions.

Equation (50) is a nonlinear one-parameter spectral problem concerning parameter c. It is shown in [7,16] that for a given even amplitude DP $F(s)$ there exist branching points of two types: eigenvalues of multiplicity two correspond to the first type, eigenvalues the multiplicity of three—to the second type. It is found in [54] analytical expressions for eigenfunctions of (50) and are obtained systems of transcendental equations for finding the possible branching points.

Using for finding the solutions of branching equation the Newton diagram method, it is shown [53] that two complex-conjugate between themselves solutions, which in the first approximation, have the form

$$f_{1,2}^{(1)}\left(s, c_1^{(l)} + \mu\right)$$
$$= f_1\left(s, c_1^{(l)}\right) + \left(a\left(s, c_1^{(l)}\right) + \alpha_{020}^{(1)}\left(s, c_1^{(l)}\right) h_1^2\right)\mu$$
$$\pm i\varphi_2\left(s, c_1^{(l)}\right) h_1 \mu^{1/2} + O\left(\mu^{3/2}\right)$$

$$\tag{51}$$

branch-off from the real solution $f_1(s)$ in the branching points of the first type $c_1^{(l)}$. Here $a\left(s, c_1^{(l)}\right)$, $\alpha_{020}^{(1)}\left(s, c_1^{(l)}\right)$ are even by s functions which are obtained by means corresponding transformations, $\varphi_2\left(s, c_1^{(l)}\right) = s \cdot f_1\left(s, c_1^{(l)}\right)$ is the second eigenfunction of (50) at the points $c_1^{(l)}$.

In [7] it is shown also that the branching-off solutions branch-off too. Analogous investigations are performed in branching points of the second type $c_2^{(l)}$.

To estimate the effectiveness of different types of solutions we consider value of the functional σ_F depending on the parameter c, which it takes on these solutions. For example, in Figure 1 are shown the values of the functional for $F(s) = 1$. The most effective solution images envelope which: on the segments I corresponds to the primary solution f_1, on II-branching-off solutions $f_{1,2}^{(1)}$ at the point $c_1^{(1)}$ with odd phase DP, on III-solutions $f_{1,2}^{(1)}$ and branching-off solutions $\tilde{f}_{1,2}^{(1)}$ from theseon IV-solutions $\tilde{f}_{1,2}^{(1)}$ and branching-off solutions $f_{1,2}^{(2)}$ at the point $c_2^{(1)}$ with even phase, на V-branching-off at the point $c_1^{(2)}$ solutions of the type $f_{1,2}^{(1)}$, on VI-solutions $f_{1,2}^{(2)}$.

Thus, the analytical investigations [7,16,52,54] and the results of numerical experiments enable to describe the general structure of the solutions of the problem dependeing on change of the value of the parameter c.

Because the values of the functional on some types of solutions in a given interval of change of the parameter c may be equal, the curves shown in Figure 1, does not map the full structure of the existing solutions. For greater clarity this structure can be represented schematically as a "tree" of solutions. In Figure 2 it is shown for the case of even DP. The primary solution $f_1(s,c)$ is central. The solution $f^{(1)}(s,c)$ with odd phase DP

$\left(\arg f^{(1)}(s,c)\right)$ branches first. At the point $\tilde{c}_1^{(1)}$ solution

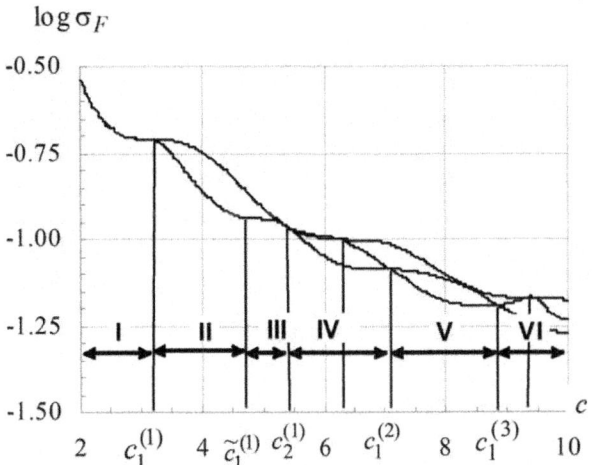

Figure 1: The value of the functional σ_F on the first primary and branching-off solutions for $F(s) = 1$.

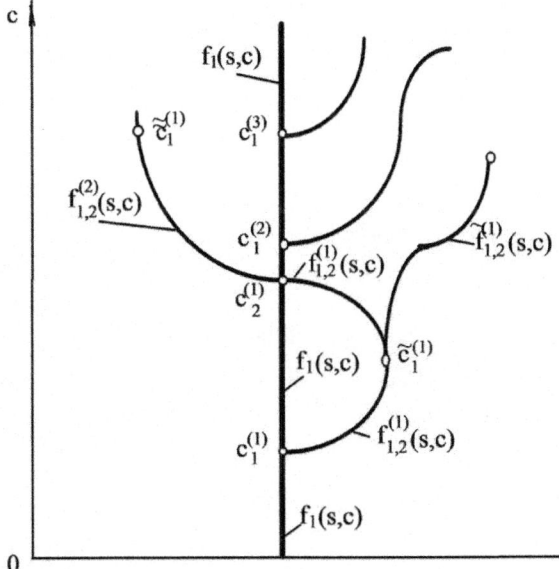

Figure 2: "Tree" of solutions.

$\bar{f}^{(1)}(s,c)$ branches from branching solution $f^{(1)}(s,c)$, and solution $f^{(1)}(s,c)$ enter in a real solution in a neighborhood of the point $c_2^{(1)}$ at c, tending to $c_2^{(1)}$. At the same point $c_2^{(1)}$ solution with even phase DP $f^{(2)}(s,c)$ branches from primary solution. The solution of the type $f^{(1)}(s,c)$ with an odd phase DP branches at the point $c_1^{(2)}$, which is located directly behind $c_2^{(1)}$. The solutions of the type $f^{(1)}(s,c)$ and the type $f^{(2)}(s,c)$ form basic branches of "tree". The possible branching points of branching-off solutions are shown on these branches.

Radiating System with A Flat Aperture

Basic Equations and Relations

Consider according with [55-58] the synthesis problems of a flat aperture assuming that form of aperture S is known and a field has elliptical polarization. Let the plane in which aperture is located, coincides with the plane XOY of the Cartesian coordinate system. Then the radiated field in the far zone can be represented by the formula [6]:

$$E(R,\vartheta,\varphi) = -\frac{k}{4\pi}\frac{e^{-ikR}}{R}D(\vartheta,\varphi)$$

where

$$D(\vartheta,\varphi) = \left[i_r \times \left[(n - i_r)\times f\right]\right],$$

$$f(\vartheta,\varphi) = \iint_S U(x,y)e^{ik(x\sin\vartheta\cos\varphi + y\sin\vartheta\sin\varphi)}dxdy$$

$$(52)$$

i_r is radial ort of spherical coordinate system, $n = -i_z$, $f(\vartheta,\varphi)$ is a vector DP of flat aperture S. Since $n = -i_z$, function $U(x,y)$ describes the tangential component of the electric vector E or vector of current flowing through the aperture S:

$$U(x,y) = U_x(x,y)i_x + U_y(x,y)i_y.$$

$$(53)$$

Introducing in a far zone special coordinate system [6] g_1, g_2, g_3 connected with orts of spherical coordinate system by formulas

$$g_1 = \cos\varphi i_\vartheta - \sin\varphi i_\varphi, \quad g_2 = \sin\varphi i_\vartheta + \cos\varphi i_\varphi,$$

$$g_3 = i_r \tag{54}$$

enables the vector synthesis problem to reduce to two independent scalar synthesis problems.

Obviously, the system g_1, g_2, g_3 is orthonormal, and transformation (54) is rotation of spherical coordinate system on an angle φ around the vector i_r. At that vector $D(\vartheta,\varphi)$ in the coordinate system (54) has the form [6]

$$D = (1+\cos\vartheta)\cdot\left[f_x g_1 + f_y g_2\right], \tag{55}$$

where

$$f_{x,y}(\vartheta,\varphi) = A_{x,y}U_{x,y}$$
$$\equiv \iint_S U_{x,y}(x,y)e^{ik(x\sin\vartheta\cos\varphi+y\sin\vartheta\sin\varphi)}dxdy. \tag{56}$$

For mappings (56) the Parseval equality [59]:

$$\left\|A_{x,y}U_{x,y}\right\|_{H_f}^2 = \left\|U_{x,y}\right\|_{H_U}^2 \tag{57}$$

are valid, that is operators A_x, A_y are isometric.

Consider the synthesis problem of a flat aperture, in which component-wise deviation of modules given and synthesized diagrams is taken into account. As optimization criterion we choose the functional type

$$\sigma_F(U)$$
$$= \iint_G \left\{\left[F_{g_1}(s_1,s_2) - \left|f_{g_1}(s_1,s_2)\right|\right]^2 \right.$$
$$\left. + \left[F_{g_2}(s_1,s_2) - \left|f_{g_2}(s_1,s_2)\right|\right]^2 \right\}ds_1 ds_2$$
$$+ \iint_{\Pi^2 G} \left\{\left|f_{g_1}(s_1,s_2)\right|^2 + \left|f_{g_2}(s_1,s_2)\right|^2\right\}ds_1 ds_2 \tag{58}$$

where F_{g_1}, F_{g_2} are modulus of components of the given amplitude DP $F = \left(F_{g_1}^2 + F_{g_2}^2\right)^{1/2}$ in closed domain

$G \in \Pi^2$. This criterion provides not only proximity of modules of given F and synthesized $|f|$ DP's, but it allows certain to influence on the polarization characteristics of the radiated field [5]. On the base of the necessary minimum condition of the functional (58) and the corresponding transformation we obtain system equations (these equations is not connected between themselves) concerning components of synthesized DP:

$$f_{g_j}(s_1,s_2) = B_j f_{g_j} \equiv \left(\frac{k}{2\pi}\right)^2$$

$$\cdot\iint_G F_{g_j}(s_1',s_2') K(s_1,s_2,s_1',s_2';k) e^{i\arg f_{g_j}(s_1',s_2')} ds_1' ds_2'$$

$$(j=1,2)$$

$$(59)$$

where

$$K(Q,Q',c) = \frac{c_1 c_2}{(2\pi)^2}$$

$$\cdot\iint_G \exp\left[i\left(c_1 x(s_1'-s_1) + c_2 y(s_2'-s_2)\right)\right] dx dy$$

$$(60)$$

is a kernel. In the case of rectangular aperture the kernel $K(s_1,s_2,s_1',s_2';c_1,c_2)$ takes the form

$$K(s_1,s_2,s_1',s_2';c_1,c_2) = \frac{\sin c_1(s_1-s_1')}{c_1(s_1-s_1')} \cdot \frac{\sin c_2(s_2-s_2')}{c_2(s_2-s_2')}$$

$$(61)$$

where

$$c_1 = ka_1 \sin \gamma_1, \; c_2 = ka_2 \sin \gamma_2,$$

$$(62)$$

are real numeric parameters characterizing the sizes of aperture a_1, a_2 in wavelengths, $k = 2\pi/\lambda$ is wave number, γ_1, γ_2 are angles that characterize the domain $\overline{G}$ (solid angle), in which different from the identity components of amplitude DP $F_{g_j}(s_1',s_2')$ are given.

Later on we omit index in (59) and shall investigate the solutions of one equation

$$f(Q) = Bf \equiv \iint_G F(Q') K(Q,Q',c) e^{i\arg f(Q')} dQ'$$

$$(63)$$

where for reduction of records we introduce the following notations

$$Q = (s_1,s_2), \; dQ' = ds_1' ds_2', \; c = (c_1,c_2).$$

Thus, the synthesis problem of flat radiating system with arbitrary polarization of irradiation according to the prescribed amplitude DP is reduced to two independent and simpler synthesis problems with linearly polarized fields in the aperture.

Equation (63) is a nonlinear two-dimensional integral equation of the Hammerstein type and it has nonunique solutions. Their quality and properties depend on the form of aperture S, the values of parameters c_1, c_2 and properties of given amplitude DP F.

On the base of decomplexification [14] we shall consider the complex space $C(\bar{G})$ as a direct sum of two real spaces of continuous functions $C(\bar{G}) = C(\bar{G}) \oplus C(\bar{G})$ in the domain $\bar{G}$. The elements of this space are written as: $f = (u,v)^{\cdot} \in C(\bar{G})$,

$u = \mathrm{Re}(f) \in C(\bar{G})$, $v = \mathrm{Im}(f) \in C(\bar{G})$. Norms in these spaces have the form:

$$\|u\|_{C(\bar{G})} = \max_{Q \in \bar{G}} |u(Q)|, \|v\|_{C(\bar{G})} = \max_{Q \in \bar{G}} |v(Q)|,$$

$$\|f\|_{C(\bar{G})} = \max\left(\|u\|_{C(\bar{G})}, \|v\|_{C(\bar{G})}\right).$$

$$(64)$$

Equation (63) in the decomplexified space $C(G)$ is reduced to equivalent to it system of equations

$$u(Q) = B_1(u,v) \equiv \iint_{\bar{G}} F(Q') K(Q,Q',c) \frac{u(Q')}{\sqrt{u^2(Q') + v^2(Q')}} dQ',$$

$$v(Q) = B_2(u,v) \equiv \iint_{\bar{G}} F(Q') K(Q,Q',c) \frac{v(Q')}{\sqrt{u^2(Q') + v^2(Q')}} dQ'. \qquad (65)$$

Denote the closed convex set of continuous functions as $S_M \subset C(\bar{G})$ setting that

$$S_M = S_{M_u} \oplus S_{M_v}, \quad S_{M_u} = \left\{ u \in S_{M_u} : \|u\|_{C(\bar{G})} \leq M \right\},$$

$$S_{M_v} = \left\{ v \in S_{M_v} : \|v\|_{C(\bar{G})} \leq M \right\},$$

$$M = \max_{Q \in \bar{G}} \iint_{\bar{G}} F(Q') |K(Q,Q',c)| dQ$$

Consider one of the properties of the function $\exp(i \arg f(Q'))$ that included in (62) at $f(Q') \to 0$. Obviously,

$$\exp\left(i\arg f\left(Q'\right)\right) = \frac{f\left(Q'\right)}{\left|f\left(Q'\right)\right|}$$

$$\equiv \frac{u\left(Q'\right) + iv\left(Q'\right)}{\left(u^2\left(Q'\right) + v^2\left(Q'\right)\right)^{1/2}} \qquad (66)$$

is a continuous function if $u\left(Q'\right) = \operatorname{Re} f\left(Q'\right)$ and

$v\left(Q'\right) = \operatorname{Im} f\left(Q'\right)$ are continuous functions, at that

$\left|\exp\left(i\arg f\left(Q'\right)\right)\right| = 1$ for any $f\left(Q'\right)$. If $u\left(Q'\right) \to 0$

and $v\left(Q'\right) \to 0$ simultaneously, then $f\left(Q'\right) \equiv 0$ is a complex zero with undefined argument by definition [60, p. 20]. On this basis at $u\left(Q'\right) \to 0$ and $v\left(Q'\right) \to 0$ we redefine $\exp\left(i\arg f\left(Q'\right)\right)$ as function, module of which is equal to one and argument is undefined.

Theorem 3.1. The operator $B = \left(B_1, B_2\right)^{\mathrm{T}}$ determined by the formulas (65) maps a closed convex set S_M of the Banach space $C\left(\bar{G}\right)$ in itself and it is completely continuous.

As the corollary from the Theorem 3.1 follows satisfaction of conditions of the Schauder principle [14, p.

411] according to which the operator $B = \left(B_1, B_2\right)^{\mathrm{T}}$ has a fixed point $f_* = \left(u_*, v_*\right)^{\mathrm{T}}$ belonging to the set S_M. This point is a solution of a system of (65) and (63), respectively.

Easily to be convinced that function

$$f_0\left(Q, c\right) = \iint_G F\left(Q'\right) K\left(Q, Q', c\right) dQ' \qquad (67)$$

is one of solutions of (63) in the case of symmetric domain $\bar{G}$.

In [56,58] it is shown that the operator

$$Df \equiv \iint_{\bar{G}} K\left(Q, Q', c\right) f\left(Q'\right) dQ'$$

is positive on the cone of nonnegative functions K of the space $C(\bar{G})$ [61]. According to this the operator D leaves invariant cone K, that is $DK \subset K$. Since $F \subset K$, the primary solution $f_0 = DF$ is also nonnegative function in the domain $\bar{G}$.

To find the branching lines and complex solutions of (63) that branch-off from the real (primary) solution $f_0(Q,c)$, we shall consider the problem on finding such a set of parameter values $c^{(0)} = \left(c_1^{(0)}, c_2^{(0)}\right)$ and all different from $f_0(Q,c)$ solutions of (65) that at

$|c - c^{(0)}| \to 0$ satisfy the conditions

$$\max_{Q \in \Omega} \left| u(Q,c) - f\left(Q, c^{(0)}\right) \right| \to 0,$$

$$\max_{Q \in \Omega} |v(Q,c)| \to 0.$$

$$(68)$$

These conditions indicate the need to find such small continuous in G solutions,

$$w(Q,c) = u(Q,c) - f_0\left(Q, c^{(0)}\right),$$

$$\omega(Q,c) = v(Q,c)$$

$$(69)$$

which converge uniformly to zero as $c \to c^{(0)}$. At that it should take into account also the direction of movement of vector c to vector $c^{(0)}$.

Set $c_1 = c_1^{(0)} + \mu$, $c_2 = c_2^{(0)} + v$, and desired solutions we find in the form

$$\begin{cases} u(Q,c) = f_0\left(Q, c^{(0)}\right) + w(Q, \mu, v), \\ v(Q,c) = \omega(Q, \mu, v). \end{cases}$$

$$(70)$$

We write the system of nonlinear integral equations of Lyapunov-Schmidt with respect to small solutions w, ω as

$$u(Q) = a_{10}\left(Q, c^{(0)}\right) \mu + a_{01}\left(Q, c^{(0)}\right) v$$

$$+ \sum_{m+n+p+q \geq 2} \mu^p v^q \iint_G A_{mnpq}\left(Q, Q', c^{(0)}\right) w^m(Q') \omega^n(Q') dQ',$$

$$(71)$$

$$\omega(Q) - \iint_G F(Q) K\left(Q, Q', c^{(0)}\right) \frac{\omega(Q')}{f_0\left(Q', c^{(0)}\right)} dQ'$$

$$= \sum_{m+n+p+q \geq 2} \mu^p v^q \iint_G B_{mnpq}\left(Q, Q', c^{(0)}\right) w^m(Q') \omega^n(Q') dQ'.$$

$$(72)$$

Here $A_{mnpq}\left(Q, Q', c^{(0)}\right)$, $B_{mnpq}\left(Q, Q', c^{(0)}\right)$ are coefficients of expansion of integrand functions of (65) in uniform convergent power series.

The problem on finding the set of possible branching points of solutions of (71) and (72) is reduced [56,58] to find the eigenvalues of two-dimensional linear homogeneous integral equation

$$\varphi(Q) = T(c_1, c_2)\varphi$$
$$\equiv \iint_\Omega \frac{F(Q')}{f_0(Q', c_1, c_2)} K(Q, Q', c_1, c_2)\varphi(Q')dQ' \tag{73}$$

at condition $f_0(Q', c) > 0$. Eigenfunctions of (73) are used [58] at the construction of branching-off solutions of (71) and (72).

Nonlinear Two-Parameter Spectral Problem

Note that (73) in the general case is a nonlinear two-parameter spectral problem. For the numerical finding the approximate solutions it is necessary to construct its digitization and consider the corresponding problem in finite-dimensional spaces. It should be noted that in the literature, in particular in [62,63], more attention is given to the construction of numerical methods for solving the nonlinear one-parameter problems.

In [64-67] a general method for finding the approximate solutions of (73), which may be applicable to a wide range of nonlinear two-parameter spectral problems is proposed.

Denote the spectral parameters as $\lambda = (\lambda_1, \lambda_2)$. Let E and V are complex Banach spaces, and the vector parameter $\lambda = (\lambda_1, \lambda_2)$ belongs to domain (open connected set) $\Lambda = \Lambda_1 \times \Lambda_2$ of the complex space $\mathbb{C}^2 = \mathbb{C} \times \mathbb{C}$, where $\lambda_i \in \Lambda_i \subset \mathbb{C}$, $(i = 1, 2)$ $\Lambda_i = \{\lambda_i \in \Lambda_i : |\lambda_i| < r_{\lambda_i}\}$, r_λ is some real constant. Consider the operator-function $A(\cdot, \cdot) : \Lambda \rightarrow L(E, V)$, where to every $\lambda = (\lambda_1, \lambda_2) \in \Lambda$ is put in correspondence operator $A(\lambda_1, \lambda_2) \in L(E, V)$.

Here the space of linear bounded operators [14] is marked as $L(E, V)$.

We shall consider the nonlinear two-parameter spectral problem of the form

$$A(\lambda_1, \lambda_2)x = 0 \tag{74}$$

where necessary to find the eigenvalues

$\lambda = (\lambda_1^{(0)}, \lambda_2^{(0)}) \in \Lambda$ and corresponding eigenvectors

$(x^{(0)} \neq 0)$ $x^{(0)} \in E$ such that $A(\lambda_1^{(0)}, \lambda_2^{(0)})x^{(0)} = 0$.

Let the Banach spaces E and E_n, $n = 1, 2, \cdots$, be given and also a system

$\mathfrak{A} = (P_n)_{n \in \mathbb{N}}$ of linear bounded operators $P_n : E \rightarrow E_n$ such that

$$\|P_n x\|_{E_n} \rightarrow \|x\|_E \quad (n \in \mathbb{N}) \; \forall x \in E .$$
(75)

Operators P_n are called connecting [14,68]. Note, by the principle of uniform boundedness [14] with (73) follows inequality $\|P_n\| \leq \text{const}$. Let in every space E_n the element x_n be selected. Writing these elements in order to increase the numbers we shall form a sequence $\{x_n\}$.

Let element x_n is selected in each space E_n. Writing these elements in ascending numerical order, we form sequence $\{x_n\}$.

Definition 3.1 [68]. The sequence $\{x_n\}_{n \in \mathbb{N}'}$ from $x_n \in E_n$ $\mathfrak{A}$-converges (discrete converges) to $x \in E$ if $\|x_n - P_n x\|_{E_n} \rightarrow 0 \; (n \in \mathbb{N}')$; we denote $x_n \xrightarrow{P} x$ $(n \in \mathbb{N}')$.

Definition of different types of convergence of operators A_n to A is given in [68]. Later on only required in further definition of stable convergence[2] A_n to A is presented.

Discretization of initial problem (74), the choice of the space E_n and definition of operators $P_n : E \rightarrow E_n$ can be differentially. In particular, one of the approaches to the digitization of (74) if the operator-function $A(\lambda_1, \lambda_2)$

is described by formula $A(\lambda_1, \lambda_2) \equiv (T(\lambda_1, \lambda_2) - I)$ where $T(\lambda_1, \lambda_2)$ is a linear continous operator and I is unique operator in the separable (infinite-dimensional) Hilbert space, consists in following. Take an arbitrary complete orthonormal system of functions $\{x_k\}_{k=1}^{\infty}$ in E. Each element $x \in E$ is represented as a series

$x = \sum_{k=1}^{\infty} c_k x_k$, where $c_k = (x, x_k)$ is Fourier coefficient of element x. Since $T(\lambda_1, \lambda_2)$ is linear continuous operator acting in separable Hilbert space, it admits the matrix representation [69]:

$$T_M (\lambda_1, \lambda_2) = \left(t_{jk} (\lambda_1, \lambda_2) \right)_{j,k=1}^{\infty} ,$$
(76)

where $t_{jk} (\lambda_1, \lambda_2) = \left(T(\lambda_1, \lambda_2) x_k, x_j \right)$. At that sequence of the Fourier coefficients of element $y = T(\lambda_1, \lambda_2) x$ is obtained from the sequence of Fourier coefficients of element x by means transformation matrix $T_M(\lambda_1,\lambda_2)$.

Using the matrix representation of the operator $T_M(\lambda_1,\lambda_2)$ in particular case (concerning Equation (73)) the spectral problem (74) is formulated as

$$A_M(\lambda_1,\lambda_2)x \equiv (T_M(\lambda_1,\lambda_2)-I_M)x=0 , \qquad (77)$$

where I_M is identity matrix in the space of sequences l_2. Thus, the operators $T(\lambda)$ and $T_M(\lambda)$ are equivalent in the sense that they put in correspondence one and the same element $y \in E$, but we obtain the Fourier coefficients of element $y=T(\lambda)x$ as a result of operation of operator $T_M(\lambda)$ on element x. Obviously, that the spectrums of these operators coincide, that is the spectral problem (77) and the problem

$$A(\lambda_1,\lambda_2)x \equiv (T(\lambda_1,\lambda_2)-I)x=0$$

are equivalent.

According to [14,68] applying to the problem (74) other discretization methods, including the following: quadrature (cubature) processes in the case of homogeneous integral equations and changing the derivatives by difference analogues in differential equations, we obtain the approximation problems for approximate finding the eigenvalues and eigenfunctions in finite-dimensional spaces

$$\varsigma(\lambda_1,\lambda_2)x_n = 0, \ n \in \mathbb{N}. \qquad (78)$$

At that the problem on finding the eigenvalues is reduced to finding the roots of the n-th order determinant, i.e. the roots of the equation

$$\Psi_n(\lambda_1,\lambda_2) \equiv \det \left\| a_{i,j}^{(n)}(\lambda_1,\lambda_2) \right\|_{i,j=1}^n = 0 \quad (n \in \mathbb{N}). \qquad (79)$$

Consider the necessary in further auxiliary one-parameter spectral problem as a particular case of (74). Set that variable λ_2 in the operator-function $A(\lambda_1,\lambda_2)$ is expressed by some unique differentiable function $\lambda_2=z(\lambda_1)$ mapping domain $\Lambda_{1,\beta} \subseteq \Lambda_1$ in some subdomain $\Lambda_{2,\beta} \subset \Lambda_2$. In the simplest case we put $\lambda_2=\beta\lambda_1$, where β is a real parameter. Introduce into consideration at $\lambda_1 \in \Lambda_{1,\beta}$ operator function $A_\beta(\lambda_1) \equiv A(\lambda_1,z(\lambda_1))$ (narrowing of operator-function $A(\lambda_1,\lambda_2)$). One-parameter nonlinear spectral problem

$$A_\beta(\lambda_1)x = 0 \qquad (80)$$

is connected with it. Here to each value

$$\lambda = (\lambda_1, z(\lambda_1)) \in \Lambda \quad \text{operator} \quad A_\beta(\lambda_1, z(\lambda_1)) \in L(E, V) \qquad \text{is} \quad \text{put} \quad \text{in}$$

correspondence.

Analogously to (78) we consider approximating sequence of discretizing problems (80) at $n \in \mathbb{N}$

$$\mathcal{A}_{\beta,n}(\lambda_1, z(\lambda_1)) x_n = 0, \quad n \in \mathbb{N}. \qquad .(81)$$

The spectrum of operator-function $A_\beta(\lambda_1)$ is denoted as $s(\Lambda_\beta)$. Suppose that $s(\Lambda_\beta) \neq \Lambda_{1,\beta}$. For spectral $s(\Lambda)$ of (74) holds [56,67].

Theorem 3.2. Let the following conditions be satisfied:

1)   operator-function $A(\cdot,\cdot): \Lambda \to L(E, V)$ is holomorphic, and $s(\Lambda) \neq \Lambda$;

2)   operator-functions $A_n(\cdot,\cdot): \Lambda \to L(E, V)$ are holomorphic and for any closed bounded set $\Lambda_0 \subset \Lambda$ the following inequality $\max\limits_{\lambda \in \Lambda_0} \|A_n(\lambda_1, \lambda_2)\| \le c(\Lambda_0) = \text{const} \ (n \in \mathbb{N})$ is valid;

3)   operators $A(\lambda_1, \lambda_2) \in L(E, V)$, $A_n(\lambda_1, \lambda_2) \in L(E_n, V_n) \ (n \in \mathbb{N})$ are the Fredholm operators with zero index for any $\lambda = (\lambda_1, \lambda_2) \in \Lambda$;

4)   spectrum $s(\Lambda_\beta) \neq \Lambda_{1,\beta}$ and a sequence of functions $\Psi_n(\lambda_1, \lambda_2)$ are differentiable in the domain Λ;

5) $A_n(\lambda) \to A(\lambda)$ is stable for any $\lambda \in r(\Lambda) = \Lambda \setminus s(\Lambda)$.

Then the following statements are true:

1)   every point of spectrum $\lambda_1^{(0)} \in s(\Lambda_\beta)$ is isolated, it is eigenvalue of the operator $A_\beta(\lambda_1) = A(\lambda_1, z(\lambda_1))$, the finite-dimensional eigensubspace $N(A(\lambda_1^{(0)}))$ and the finite-dimensional root subspace correspond to it;

2)   for each $\lambda_1^{(0)} \in s(\Lambda_\beta)$ there exists a sequence

$\left\{\lambda_{1,n}^{(0)}\right\}$ from $\lambda_{1,n}^{(0)} \in s\left(A_{\beta,n}\right)$ $\left(n > n_0\right)$, such that

$\lambda_{1,n}^{(0)} \to \lambda_1^0$;

3)   each point $\lambda^{(0)} = \left(\lambda_1^{(0)}, z\left(\lambda_1^{(0)}\right)\right) \in \Lambda$ is a spectrum point of the operator-function $A\left(\lambda_1, \lambda_2\right)$;

4)   if in some small ε_0 -neighborhood of the point $\lambda^{(0)} = \left(\lambda_1^{(0)}, z\left(\lambda_1^{(0)}\right)\right) \in \Lambda$ at all n larger any number N_0

(corresponding ε_0, according to definition of limit of sequence p. 2)) the sequence of partial derivates

$\left\{\dfrac{\partial \Psi_n}{\partial \lambda_2}\left(\lambda_{1,n}^0, z\left(\lambda_{1,n}^0\right)\right)\right\}$ is nonzero, then in an arbitrarily small ε_* -neighborhood of point $\left(\lambda_1^{(0)}, z\left(\lambda_1^{(0)}\right)\right) \in \Lambda$ there exists a continuous differentiable function

$\lambda_{2,N_*} = \varphi_{N_*}\left(\lambda_1\right)$, which is solution of (89), at that

$\lambda_{2,N_*}^{(0)} = \varphi_{N_*}\left(\lambda_{1,N_*}^{(0)}\right)$ and at the point

$\left(\lambda_{1,N_*}^{(0)}, \lambda_{2,N_*}^{(0)}\right) = \left(\lambda_{1,N_*}^{(0)}, \varphi_{N_*}\left(\lambda_{1,N_*}^{(0)}\right)\right)$ however little differs from point of a spectrum of auxiliary one-parameter problem (91) $\left|\lambda_{1,N_*}^{(0)} - \lambda_1^{(0)}\right| < \varepsilon_*$; that is in some bicylindrical domain

$\Lambda_0 = \left\{\left(\lambda_1, \lambda_2\right) \in \Lambda_0 : \left|\lambda_1 - \lambda_1^{(0)}\right| < \varepsilon_1, \left|\lambda_2 - \lambda_2^{(0)}\right| < \varepsilon_2\right\}$ there exists a connected component of spectrum of the operator-function $A_{N_*}\left(\lambda_1, \lambda_2\right)$ ($\varepsilon_1, \varepsilon_2$ are small real constants).

Proof. The proof of the theorem is given in [56] and is based on Theorems 1 and 2 with [68, pp. 68-69] and the Theorem about existence of implicit function (see, for example, [70]).

If the points $\left(\lambda_{1,\nu}^{(0)}, \lambda_{2,\nu}^{(0)}\right) \in \Lambda$ are the eigenvalues of

(78) and derivatives $\partial \Psi_n / \partial \lambda_1$, $\partial \Psi_n / \partial \lambda_2$ in these points are nonzero, to find connected components of the spectrum of this problem on the base of (79) Cauchy problem [56,57,65] we solve the in a neighborhood of each point

$$\left(\lambda_{1,\nu}^{(0)},\lambda_{2,\nu}^{(0)}\right)\in\Lambda$$

$$\frac{d\lambda_2}{d\lambda_1}=-\frac{\partial\Psi_n\left(\lambda_1,\lambda_2\right)/\partial\lambda_1}{\partial\Psi_n\left(\lambda_1,\lambda_2\right)/\partial\lambda_2},$$

(82)

$$\lambda_2\left(\lambda_{1,\nu}^{(0)}\right)=\lambda_{2,\nu}^{(0)}.$$

(83)

Numerical Algorithms for Finding the Possible Branching Lines of Solutions

Return to finding the solutions of (73), in which c_1, c_2 are spectral parameters. Let $(c_1,c_2)\in\Lambda_c$,

$\Lambda_c=\Lambda_{c_1}\times\Lambda_{c_2}$, where $\Lambda_{c_i}=\left\{c_i\in\Lambda_{c_i}:0<c_i<r_c\right\}$. By direct check we set that for arbitrary values of the parameters $(c_1,c_2)\in\Lambda_c$ the function

$$\varphi_0(Q,c)=\iint\limits_{\Omega}F(Q')K(Q,Q',c)dQ'=f_0(Q,c)$$

(84)

is one of the eigenfunctions, that is there exists a connected set of the spectrum, coinciding with the domain Λ_c. As a result of this, the condition $s(\tilde{\Lambda})\neq\Lambda_c$ is not satisfied. To find another connected components of spectrum we exclude eigenfunction (73) from the kernel of integral equation, namely, consider the equation

$$\varphi(Q,c)=\tilde{T}(c)\varphi\equiv\iint\limits_{\Omega}\tilde{K}(Q,Q',c)\varphi(Q')dQ',$$

(85)

where

$$\tilde{K}(Q,Q',c)=\frac{F(Q')}{f_0(Q',c)}K(Q,Q',c)$$
$$-\frac{\psi_0(Q)\varphi_0(Q',c)}{\|\psi_0\|\|\varphi_0\|}.$$

(86)

Here $\psi_0(Q)$ is adjoint with (73) eigenfunction of equation of From Lemma Schmidt [53, p. 132] follows that from spectrum of operator $\tilde{T}(c)$ is excluded coherent component coinciding with the domain Λ_c and the corresponding to the function $\varphi_0(Q,c)$.

Using to (73) certain convergent cubature process with coefficients $a_{jn} \in \mathbb{R}$ and nodes $Q_{jn} \in \bar{\Omega}$ $(n \in \mathbb{N})$ and rejecting in it remainder, we obtain homogeneous system of linear algebraic equations (SLAE)

$$
\begin{aligned}
u_{in} &= T_{M_n}(c_1, c_2) u \\
&\equiv \sum_{j=1}^{n} a_{jn} \tilde{K}(Q_{in}, Q_{jn}, c_1, c_2) u_{jn} \quad (i = 1 \div n)
\end{aligned}
\tag{87}
$$

where $u_{in} = u(Q_{in})$.

The presence of such values of parameters c_1, c_2, which are the solutions of the equation

$$
\Psi_n(c_1, c_2) = \det\left(T_{M_n}(c_1, c_2) - I_n\right) = 0
\tag{88}
$$

is necessary condition of the existence different from zero solutions of (87). We consider (88) as the problem on finding the implicitly given function $c_2 = \gamma(c_1)$, reducing it to the Cauchy problem (82) and (83). Putting $c_2 = \beta c_1$ in (85), we shall consider the auxiliary one parameter spectral problem

$$
\varphi(Q) = \tilde{T}(c_1)\varphi \equiv \iint_{\Omega} \tilde{K}(Q, Q', c_1)\varphi(Q')dQ'
$$

solutions of which we use as initial conditions in the Cauchy problem (83). Corresponding this equation SLAE has the form

$$
\begin{aligned}
u_{in} &= \tilde{T}_{M_n}(c_1) u_n \\
&\equiv \sum_{j=1}^{n} a_{jn} \tilde{K}(Q_{in}, Q_{jn}, c_1) u_{jn} \quad (i = 1, \cdots, n)
\end{aligned}
\tag{89}
$$

and the problem on finding the eigenvalues of this system is reduced to finding the roots of the equation

$$
\tilde{\Psi}_n(c_1) = \det\left(\tilde{T}_{M_n}(c_1) - I_n\right) = 0
$$

. For the numerical solution of the Cauchy problem (82) and (83) are used the Runge-Kutta and Adams methods.

We shall present numerical examples of finding the solutions of (73) for two given amplitude DP's. In Figures 3 and 4 are shown spectral lines of (73), corresponding the given DP $F(s_1, s_2) \equiv 1$ and given DP which is defined by the formula:

$$F\left(s_1, s_2\right) = \begin{cases} 2\sqrt{\left(s_1^2 + s_2^2\right)} \cdot \sqrt{\left(1 - \left(s_1^2 + s_2^2\right)\right)}, & \left(s_1^2 + s_2^2\right) \le 1, \\ 0, & \left(s_1^2 + s_2^2\right) > 1. \end{cases}$$

(90)

Note that to each point of the spectral lines given in these figures correspond the eigenfunctions of (73) with the characteristic properties for each line. For example, below are shown the eigenfunctions that correspond to points of intersection of the spectral lines 1 and 2 (Figures 3 and 4) with the beam $c_2 = 0.8c_1$.

Variational Approach to Solution of the Nonlinear Spectral Problems

In [71,72] along with the implicit functions method a variational approach to solution of the nonlinear oneparameter and two-parameter spectral problems on finding the eigenvalues $\lambda = \{\lambda_1, \lambda_2\} \in \Lambda_c \subset \mathbb{R}^2$ and eigenelements $u \in U \subset L_2(\Omega)$ of equation

$$T(\lambda)u = u$$

(91)

in the real Hilbert space $L_2(\Omega)$ for the case when $T(\lambda): L_2(\Omega) \to L_2(\Omega)$ is a linear positive definite

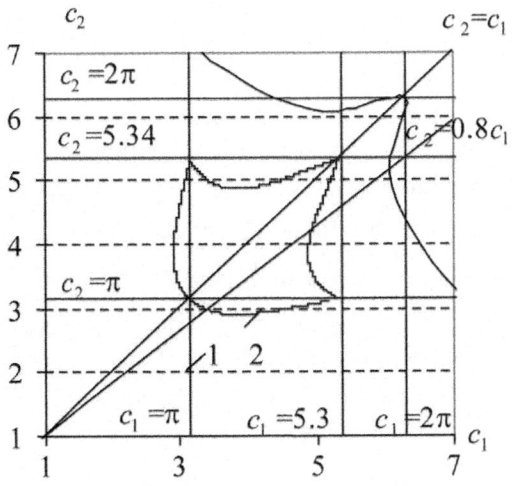

Figure 3: The possible branching lines of solutions of system of (65) for $F(s_1, s_2) = 1$.

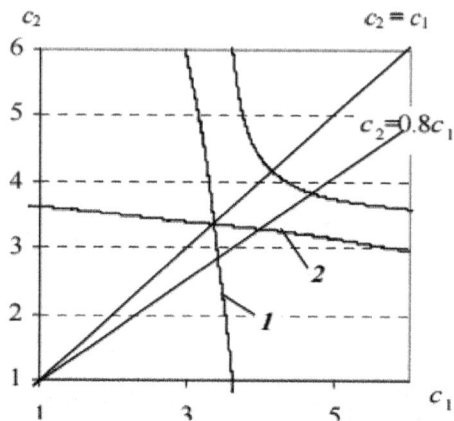

Figure 4: The possible branching lines of solutions of system of (65) for $F(s_1,s_2)$, which is defined by (90).

self-adjoint operator nonlinearly depending on the parameters λ_1,λ_2, is proposed. Variational problem is formulated as the problem on finding such values of parameter $\lambda \in \Lambda_c$ and such functions $u_* \in L_2(\Omega)$ on which functional

$$\varphi(\lambda,u) = \left\| T(\lambda)u - u \right\|_{L_2}^2$$
$$\equiv \left(T(\lambda)u - u, T(\lambda)u - u \right) \tag{92}$$

becomes minimum. The equivalence of the spectral problem (91) and put it in correspodence of variational problem (92) is proved. Based on the method of generalized coordinate descent iterative process for the numerical finding one of the eigenvalues and the corresponding eigenfunction of (91) is suggested. Local convergence is proved.

Example of use of a variational approach to finding the eigenvalues and eigenfunctions of (73) is shown in Figure 5 for $F(s_1,s_2) \equiv 1$ and in Figure 6 for the case when the function $F(s_1,s_2)$ is defined by (90). Later on the eigenfunctions of (73), corresponding to eigenvalues belonging to curves 1, 3 illustrated in Figure 5(b) are shown. From the analysis of the figures we see that the eigenfunctions $\varphi_1(s_1,s_2)$ are odd by argument s_1 and functions $\varphi_2(s_1,s_2)$ are odd by both arguments.

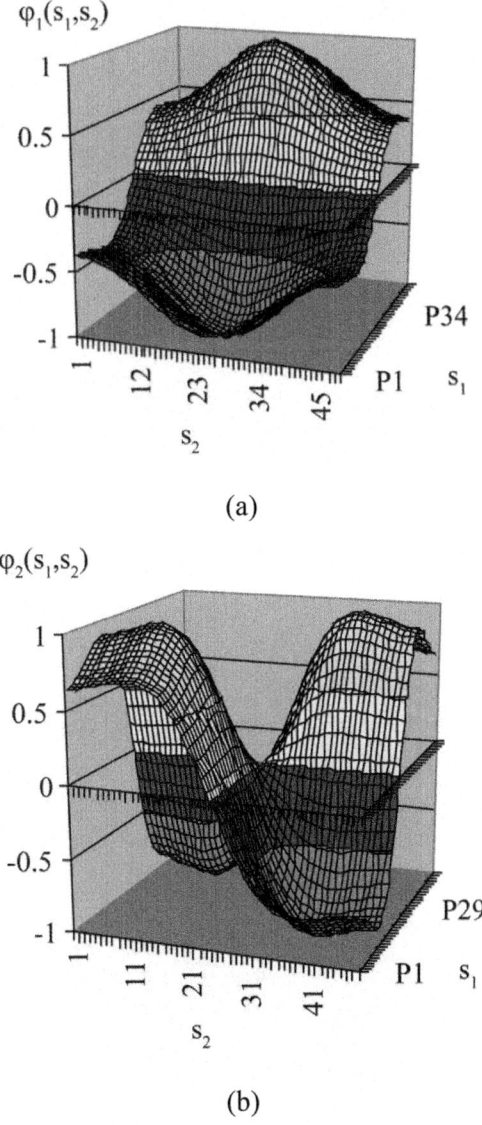

(a)

(b)

Figure 5: Normalized eigenfunctions of (73) corresponding to the eigenvalues: (a) (c_1 = 3.14159393, c_2 = 2.51327514); (b) (c_1 = 3.64391021, c_2 = 2.91512817).

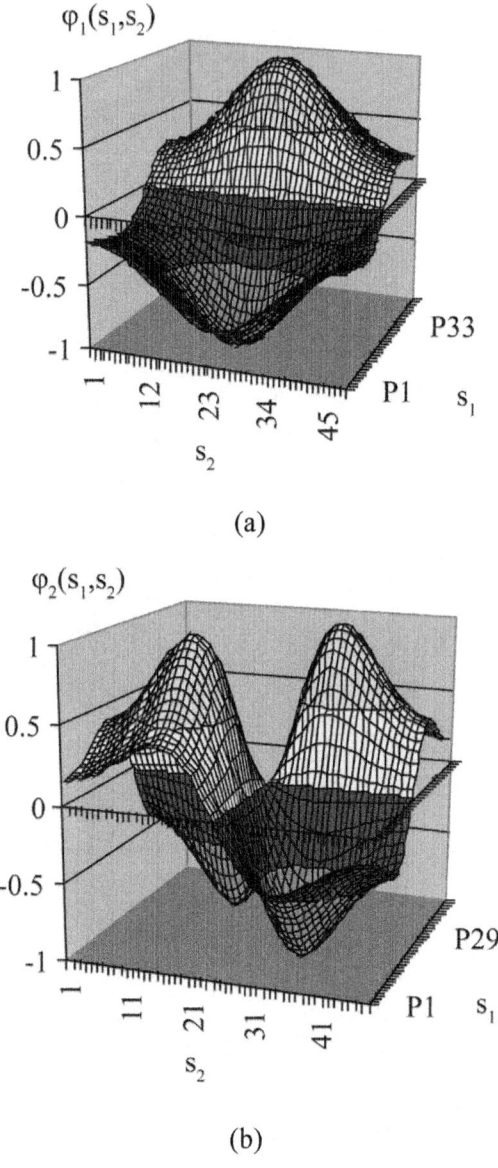

(a)

(b)

Figure 6: Normalized eigenfunctions of (73) corresponding to the eigenvalues: (a) ($c_1 = 3.43408813$, $c_2 = 2.74727050$); (b) ($c_1 = 4.18890991$, $c_2 = 3.351127928$).

Found by numerical method form and properties of eigenfunctions in the possible branching points are used to determine of the properties of branching-off in these points of solutions of nonlinear systems of (65).

About Branching of Solutions in the Case of a Flat Aperture

In [56,58,73,74] using the found branching lines and eigenfunctions, the analytical investigations of branching of the primary solution of the first type of (65) for the case when the the the kernel $K\left(s_1,s_2,s_1',s_2';c_1,c_2\right)$ has the form (61), and the multiplicity of eigenvalues of the linear Equation (73) at the branching points $\left(c_1^{(0)},c_2^{(0)}\right)$ is two, are presented.

The study of solutions of (65) is realized on the beam $c_2=\beta c_1$ belonging to the domain Λ_c. Let

$$c^{(0)}=\left(c_1^{(0)},c_2^{(0)}\right)=\left(c_1^{(0)},\beta c_1^{(0)}\right)$$ be eigenvalue of (73). We assign to parameter $c_1^{(l)}$ the small disturbance

$c_1=c_1^{(0)}+\mu$, $c_2=\beta c_1^{(0)}+\beta\mu$ and consider the problem on finding all different from $f_0\left(s_1,s_2,c_1,c_2\right)$ solutions of (65), which at $\mu\to 0$ satisfy the conditions

$$\max_{Q\in\Omega}\left|u\left(Q,c_1,c_2\right)-f_0\left(Q,c_1^{(0)},\beta c_1^{(0)}\right)\right|\to 0$$
,

$$\max_{Q\in\Omega}\left|v\left(Q,c_1,c_2\right)\right|\to 0$$
.

The system of (65) by means of expanding the integrand functions is reduced to the corresponding system of Lyapunov-Schmidt equations, similar to (71) and (72). Desired solutions are found in the form

$$u\left(Q,c_1,c_2\right)=f_0\left(Q,c_1^{(0)},\beta c_1^{(0)}\right)+w\left(Q,\mu\right)$$
,

$$v\left(Q,c_1,c_2\right)=\omega\left(Q,\mu\right)$$
.

As a result we obtain [74] that at the points $\left(c_1^{(0)},c_2^{(0)}\right)=\left(c_1^{(0)},\beta c_1^{(0)}\right)$ from the primary solution $f_0\left(s_1,s_2,c_1^{(0)},\beta c_1^{(0)}\right)$ branch-off two complex-conjugate solutions having in the first approximation the form

$$f_{1,2}^{(1)}\left(s_1,s_2,c_1,\beta c_1\right)$$

$$= f_0\left(s_1,s_2,c_1^{(0)},\beta c_1^{(0)}\right)+\left[a\left(s_1,s_2,c_1^{(0)},\beta c_1^{(0)}\right)\right.$$

$$\left.+\alpha_{020}^{(1)}\left(s_1,s_2,c_1^{(0)},\beta c_1^{(0)}\right)h_1^2\right]\mu$$

$$\pm i\,\frac{\varphi_1\left(s_1,s_2,c_1^{(0)},\beta c_1^{(0)}\right)}{\left\|\varphi_1\left(s_1,s_2,c_1^{(0)},\beta c_1^{(0)}\right)\right\|}\,h_1\mu^{1/2}+O\left(\mu^{3/2}\right).$$

$$(93)$$

The imaginary part being determined by the properties of eigenfunctions $\varphi_1\left(s_1,s_2,c_1^{(0)},\beta c_1^{(0)}\right)$. Functions

$\arg f_{1,2}^{(1)}\left(s_1,s_2,c_1,\beta c_1\right)$, obtained on the base of (93), determine the properties of the phase DP and APD of the field in aperture. The properties obtained in the first approximation of solutions agree with numerical results.

For example, in Figure 7 are shown the values of the functional σ_F at $F\left(s_1,s_2\right)\equiv 1$, which it takes on the primary (curve 1) and branching-off (curves 2, 3, 4) solutions on the beam $c_2=0.8c_1$. Note, that on the segment $\left(c_1^{(1)},c_1^{(5)}\right)$ the branching-off solutions with an odd phase DP $\arg f\left(s_1,s_2\right)$, to which the nonsymmetric amplitude-phase distribution of the field in aperture corresponds, are the most effective. On the segment $\left(c_1^{(5)},8\right)$ the most effective is the solution of 4 with properties $\arg f_{1,2}^{(1)}\left(-s_1,s_2\right)=\arg f_{1,2}^{(1)}\left(s_1,s_2\right)$,

$\arg f_{1,2}^{(1)}\left(s_1,-s_2\right)=\arg f_{1,2}^{(1)}\left(s_1,s_2\right)$. The symmetric but complex APD of the field in aperture corresponds to it. From the analysis of Figure 7 follows that the same efficiency of the synthesis can be achieved on the branching-off solutions at smaller sizes of aperture and smaller values of parameters c_1, c_2, than on the primary solution. The linear size of aperture can be decreased by the amount $\delta_1 c_1$ or $\delta_2 c_1$ at realization of branching-off solution.

Numerical examples of synthesis of given funnelshaped amplitude DP defined in the domain G by (90), are given in Figures 8 and 9. The branching lines of solutions of (63) for this DP are shown in Figure 4. The given DP and optimum synthesized DP are presented in Figures 8(a) and (b), respectively, at $c_1=9.25$, $c_2=7.4$. The optimum amplitude distribution of the field in an aperture $U\left(x,y\right)$, which creates given in Figure 8(b) the synthesized DP, is shown in Figure 9. From the analysis of these figures we see that the symmetric amplitude DP (Figure 8(b)) can be created by different distributions of the field in aperture of radiating system, including real and nonsymmetric distribution (Figure 9).

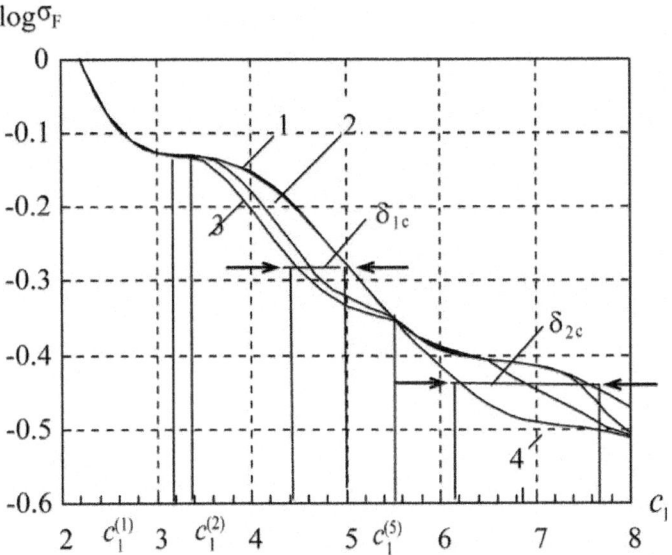

Figure 7: The values of the functional σ_F at $F(s_1,s_2) \equiv 1$.

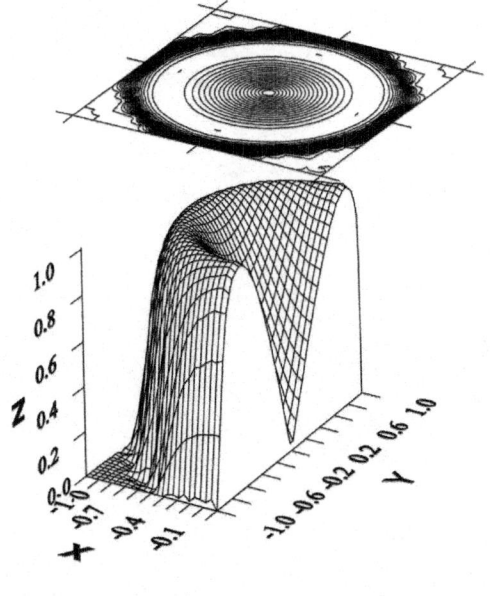

(a)

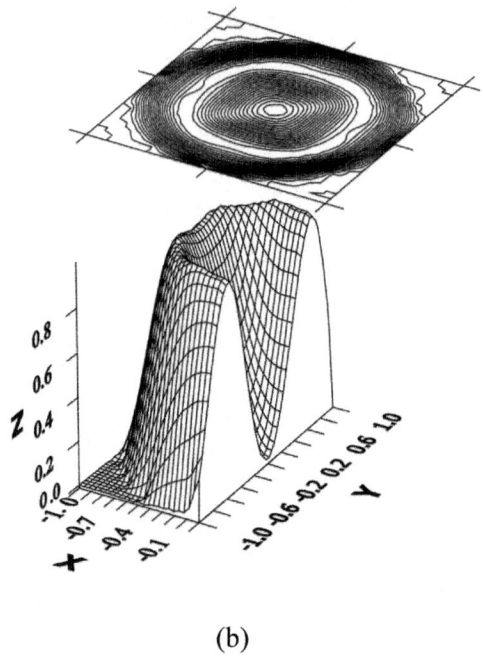

(b)

Figure 8: The prescribed (a) and synthesized (b) DP.

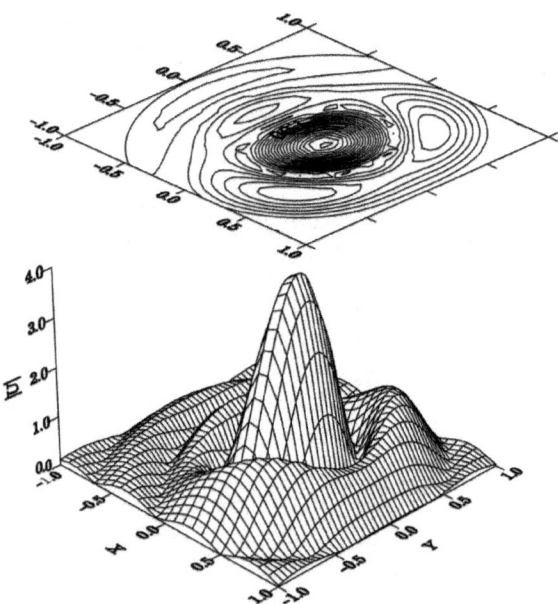

Figure 9: Amplitude distribution of field in aperture which creates DP given in **Figure 8**(b).

SYNTHESIS OF DISCRETE RADIATING SYSTEMS— ANTENNA ARRAYS (AA)

The investigations of nonlinear synthesis problems of linear and planar antenna arrays (AA) according to the prescribed amplitude DP are presented partially in [7,24, 28-37]. In the basis of construction of mathematical models it is assumed [6,27] that the excitation of each radiator is characterized by a single complex number I_n - complex amplitude of excitation, the physical meaning of which depends on the type of radiating system. Taking into account the linearity of Maxwell's equations, the complex amplitudes of excitation enter linearly in the expression for DP of array, that is

$$f(\theta,\varphi)=\sum_{n=1}^{N} I_n f_n(\theta,\varphi) e^{ik(x_n \sin\theta\cos\varphi + y_n \sin\theta\sin\varphi + z_n \cos\theta)}$$

$$(94)$$

Here $f_n(\theta,\varphi) = f_\theta^{(n)}(\theta,\varphi) i_\theta + f_\varphi^{(n)}(\theta,\varphi) i_\varphi$ is a vector DP of n-th radiator. Vector $I = \{I_1, I_2, \cdots, I_N\}$ is called the vector of excitation of array or vector of amplitude-phase distribution of currents in the array. Such formulation of DP of array is used in the synthesis problems with regard for mutual influence of radiators [27, 29,30]. Thus the problem on finding the functions $f_n(\theta,\varphi)$ is reduced to solution of the corresponding boundary problem of electrodynamics in multiply connected domains [2,4,39,40]. The method of integral equations [40,42] is used widely in such classes of problems. The synthesis method of antenna arrays with cylindrical dipoles with account of mutual influence is proposed in [29,34]. Analysis of nonuniqueness problem of solutions is studied there by means of computational experiments.

In the problems of analysis and synthesis of antenna arrays with many elements is used simplified mathematical model of AA [5,6]. It is assumed [6] that AA consists of N identical and identically oriented in space radiators, and vector DP of radiators are identical for all emitters, i.e. $(n=1\div N)$ $f_n(\theta,\varphi) = f^{(R)}(\theta,\varphi)$. Formula (94) for DP of flat AA takes the form

$$f(\theta,\varphi)$$
$$= f^{(R)}(\theta,\varphi) f_\Sigma(\theta,\varphi)$$
$$\equiv f^{(R)}(\theta,\varphi) \sum_{n=-M_1}^{M_1} \sum_{m=1}^{N(n)} I_{nm} e^{i(c_1 x_{nm} s_1 + c_2 y_{nm} s_2)}$$

$$(95)$$

Here $s_1 = (\sin\theta\cos\varphi)/\sin\gamma_1$, $s_2 = (\sin\theta\sin\varphi)/\sin\gamma_2$ are the generalized angular coordinates,

$$c_1 = kd_x \sin\gamma_1, \quad c_2 = kd_y \sin\gamma_2$$

are dimensionless numerical parameters characterizing the distance between the radiators and the domain (solid angle) G, in which the required amplitude DP $F(s_1,s_2)$ is given. Since in (95) only the second multiplier depends on the vector APD of excitation currents in the array:

$$f_\Sigma(\theta,\varphi) = AI \equiv \sum_{n=-M_1}^{M_1} \sum_{m=1}^{N(n)} I_{nm} e^{i(c_1 x_{nm} s_1 + c_2 y_{nm} s_2)}, \tag{96}$$

only the synthesis problem of factor of AA is considerd. Function $f_\Sigma(s_1,s_2)$ is $2\pi/c_1$-periodic by argument s_1 and $2\pi/c_2$-periodic by s_2. We consider also (96) as the action of the operator A from a finite-dimensional space $H_i = \mathbb{C}^N$ (N is number of radiators) into the finite-dimensional subspace of the space $H_f = C(\bar\Omega_p)$ where $\bar\Omega_p$ is the domain corresponding to the period of array. Let the amplitude DP $F(s_1,s_2)$ be given in the domain $\bar G \subset \bar\Omega_p$, and on the set $\bar\Omega_p \setminus \bar G$ is identically equal to zero. The synthesis problem is to minimize the functional [35]

$$\min_{I \in H_I} \sigma_F(I) = \|F - |AI|\|^2_{C^{(2)}_{(\bar\Omega_p)}} \equiv \|F - |f|\|^2_{C^{(2)}_{(\bar\Omega_p)}}. \tag{97}$$

The basic of synthesis equations of multiplier of AA have the form

$$I = A^*(Fe^{i\arg AI}) \tag{98}$$

the equation concerning APD of currents in AA, where A^* is conjugate with A operator, and

$$f(Q) = Bf \equiv \iint_{\bar G} K_{ar}(Q,Q',c)F(Q')e^{i\arg f(Q')}dQ' \tag{99}$$

is equation concerning of synthesized DP. Here

$Q = (s_1,s_2)$, $dQ = ds_1 ds_2$, $c = (c_1,c_2)$; $K_{ar}(Q,Q',c)$ is the kernel the form of which depends on the distribution of elements in AA. In particular, in the case of a rectangular array with number of elements

$N_1 \cdot N_2 = (2M_1+1)\cdot(2M_2+1)$ the kernel $K_{ar}(Q,Q',c)$ is written as

$$K_{ar}(Q,Q',c_1,c_2)$$

$$= \frac{\sin\left(N_1 \frac{c_1}{2}(s_1 - s_1')\right)}{\sin\left(\frac{c_1}{2}(s_1 - s_1')\right)} \cdot \frac{\sin\left(N_2 \frac{c_2}{2}(s_2 - s_2')\right)}{\sin\left(\frac{c_2}{2}(s_2 - s_2')\right)}.$$

(100)

To find the possible branching lines of solutions of (99) a linear homogeneous integral equation

$$\varphi(Q) = T(c_1,c_2)\varphi$$
$$= \iint_G F(Q')K_{ar}(Q,Q',c_1,c_2)/f_0(Q',c_1,c_2)\varphi(Q')dQ'$$

(101)

is obtained where $f_0(Q',c_1,c_2)$ is a primary solution of (99).

Note that the kernel $K_{ar}(Q,Q',c_1,c_2)$ is degenerate. Consequently, Equation (101) is reduced to the corresponding homogeneous SLAE what in a special case of rectangular array has the form

$$x_{kl} = \sum_{m=-M_1}^{M_1} \sum_{n=-M_2}^{M_2} a_{nm}^{(kl)}(c_1,c_2) x_{nm}$$
$$(k = -M_1 \div M_1, l = -M_2 \div M_2)$$

(102)

Coefficients of this system depend nonlinearly on the spectral parameters c_1,c_2 and on the given amplitude DP. In [35,65] the conditions are determined and the existence theorem of connected components of the spectrum of (101) is proved. To find the spectral lines the implicit function method (82) and (83), is used. Consider the numerical results of finding the solutions of the branching lines in the synthesis problems of a plane equidistant antenna array with 11×11 radiators for two given in the domain $\bar{G} = \{(s_1,s_2): |s_1| \leq 1, |s_2| \leq 1\}$

amplitude   DPs $F(s_1,s_2) = \cos(\pi s_1 / 2)|\sin(\pi s_2)|$ (Figure   10)   and $F(s_1,s_2) = |\sin(\pi s_1)| \cdot |\sin(\pi s_2)|$ (Figure 11), which are obtained by solving of (101) and (102).

The prescribed and synthesized amplitude DPs (with phase DP odd by argument s_2) at $c_1 = 1.25$, $c_2 = 1.125$, are shown in Figure 12 and 13, respectively. The amplitude and phase distributions of currents in the array of corresponding synthesized DP are given in Figure 14. From the analysis of this figure we see that nonsymmetric Y-direction distribution of currents in the array forms symmetrical amplitude DP.

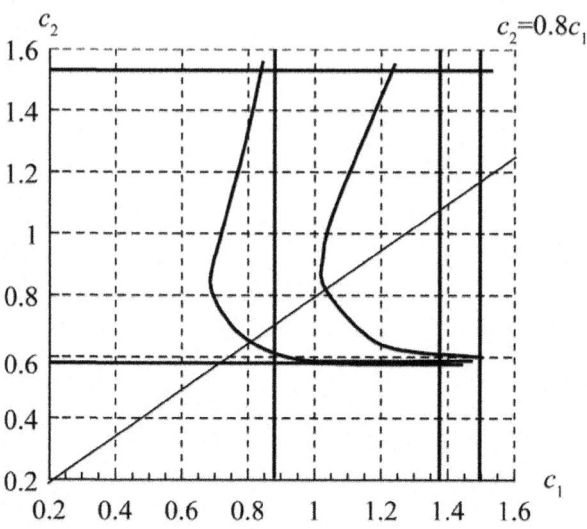

Figure 10: The possible branching lines of solutions of (99) for $F(s_1,s_2) = \cos(\pi s_1/2)|\sin(\pi s_2)|$.

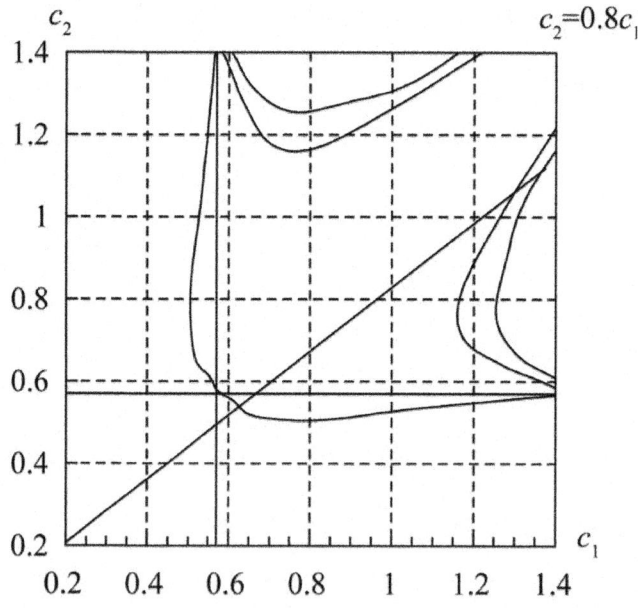

Figure 11: The possible branching lines of solutions of (99) for $F(s_1,s_2) = |\sin(\pi s_1)||\sin(\pi s_2)|$

.

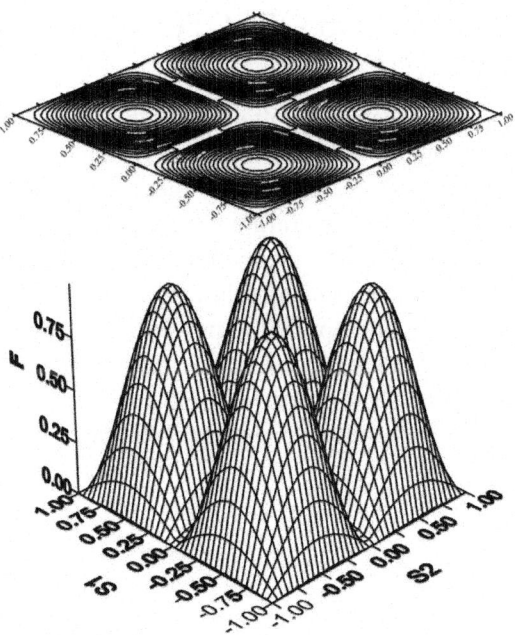

Figure 12: The prescribed DP.

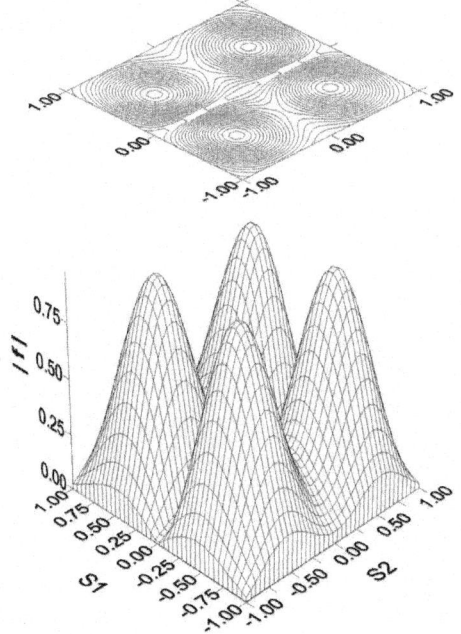

Figure 13: The synthesized DP.

NUMERICAL METHODS OF SOLUTION OF THE BASIC SYNTHESIS EQUATIONS

The above results show that the nonlinear synthesis problems according to the prescribed amplitude DP and given energy DP have nonunique solutions. Application of the methods of branching theory of solutions to nonlinear integral equations allows to determine the quantity of existing solutions, to find solutions in the first approximation and to determine their quality characteristics. To find the complete solutions of these equations numerical methods [7,29,36,49,75] are applied. The defined properties of solutions obtained by analytical investigations make it possible to choose the initial approximation having the basic properties of the desired solutions and they are placed in certain neighborhoods of complete solutions.

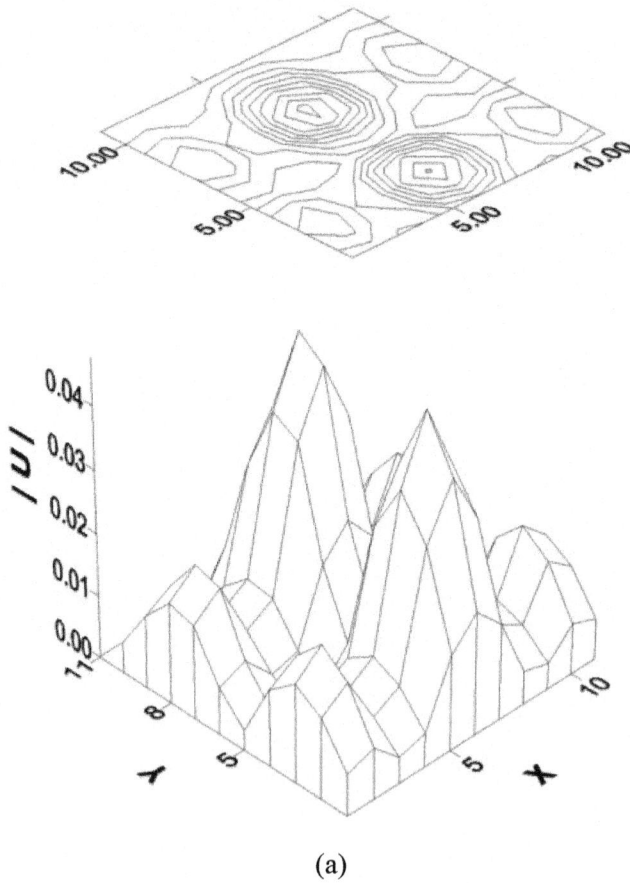

(a)

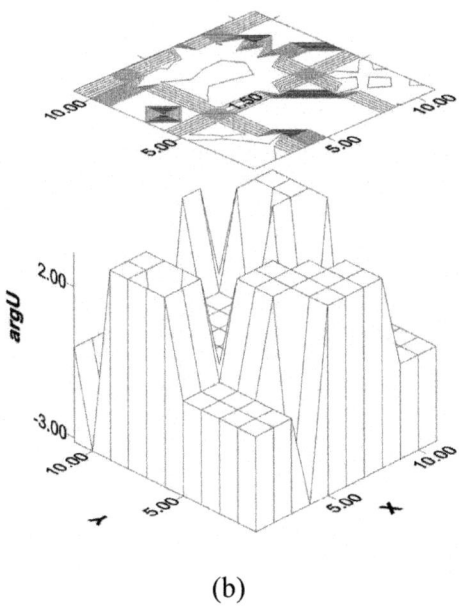

(b)

Figure 14: The optimal amplitude (a) and phase (b) distributions of currents in the array.

Conditionally the process of numerical solution of synthesis problem can be divided into two stages. The first of them is described partially above and it consists in finding the points (lines) of branching and determination of types of existings solutions depending on the value of physical parameters. The second stage consists in solving the basic synthesis equations by iterative methods.

Numerical Solution of Synthesis Equations Corresponding to Functional σ_F

As an example of the scalar problem we consider iterative process of solving the equation of type (9), in the base of which we put the successive approximations method [7,75]

$$f_n = AA^* \left(F e^{i \arg f_{n-1}} \right) \quad (n=1,2,3,\cdots) \tag{103}$$

Obviously, the successive approximations method (103) is equivalent to the following iterative process

$$\begin{cases} U_n = A^* \left(F e^{i \arg f_{n-1}} \right), \\ f_n = AU_n \quad (n=1,2,3,\cdots). \end{cases} \tag{104}$$

In [75] it is shown that the sequences $\{U_n\}$ and $\{f_n\}$ generated by iterative process (104), are relaxational for functional σ_F. Relaxation properties of (104) states Theorem 5.1. The sequence $\{U_n\}$ is generated by the iterative process (104), it is relaxation for functional $\sigma_F(U)$, and the values which it takes on $\{U_n\}$ form a convergent numerical sequence $\{\sigma_F(U_n)\}$.

Formulate also the properties of the operator B entering in (41) that complement the properties of $1°$ - $3°$ solutions of (41), presented in Section 3.1.

Theorem 5.2. Nonlinear operator B, defined by (41), acts in the space $C[-1,1]$ of continuous complex- valued functions, it is a compact and maps set $S_M = \{f : \|f\|_C \le M\}$ it into itself, where

$$M = \max_{s\in[-1,1]} \int_{-1}^{1} F(s')|K(s,s';c)|\,ds'$$

(105)

that is $B(S_M) \subset S_M$.

From the proved theorem follows, in particular, the following fact. Since the solutions of (41) are fixed points of the operator B, from the relation $B(S_M) \subset S_M$ follows that all solutions of this equation belong to the set $B(S_M) \subset S_M$. In addition, is valid [75]

Corollary 5.1. If the sequence $\{f_n\}$ which is generated by the iterative process (104), is minimizing for the functional $\sigma_F(U)$, then from $\{f_n\}$ can be selected a subsequence $\{f_{n_k}\}$ converging uniformly to the minimum point f_* of the functional $\sigma_F(U)$.

Note that Theorem 5.2 and Corollary 5.1 are extended to the case of synthesis problem of a flat aperture with use of equation of the type (63).

Numerical Solution of Synthesis Equations Corresponding to Functional σ_{F_α}

In the base of construction of iterative processes of solving the nonlinear operator equations of the type (12) and (13) we put implicit scheme of the successive approximations method [7,76]. In a general case, the iterative process of solution of (12) has the form

$$\left(E+\alpha^{-1}A^*A\right)U_{n+1}$$
$$=\alpha^{-1}A^*\left(F\exp\left(i\arg\left(AU_n\right)\right)\right)\quad\left(n=0,1,2,\cdots\right), \tag{106}$$

where E is an identity operator acting in the space $H_U=L_2(V)$.

The implicit scheme of iteration process for (13) with respect to synthesized DP f [3] has the form similar to (106)

$$\left(E+\alpha^{-1}AA^*\right)f_{n+1}$$
$$=\alpha^{-1}AA^*\left(F\exp\left(i\arg\left(f_n\right)\right)\right)\quad\left(n=0,1,2,\cdots\right). \tag{107}$$

Note that the implicit schemes (106) and (107) are characterized by the fact that linear operator equation is solved on every iteration step. In addition the question of solvability of (106) and (107) appears, to which a positive answer gives a theorem about the solvability of the functional equation of the second kind of the type

$$x=Wx+y \tag{108}$$

in the Banach space X, where W is a linear compact operator [69].

Theorem 5.3 [69]. In order that (108) have the solution at an arbitrary $y\in X$, it is necessary and sufficiently that homogeneous equation $x=Wx$ have a unique solution (obviously, that $x=0$).

For a sequence $\{U_n\}$ obtained by (106), is valid Theorem 5.4. Let $A:L_2(\bar V)\to L_2(\bar G)$ be a completely continuous operator, F be a continuous real nonnegative function in $\bar G$ and at $0<\alpha<\infty$ there exists the inverse operator $\left(E+\alpha^{-1}A^*A\right)^{-1}$, in additionthe dimension of the space of zeros $N(A)=0$.

Then the sequence $\{U_n\}$ generated by the iterative process (106), is a minimizing for the functional

$$\left\|\mathrm{grad}\sigma_{F_\alpha}\left(U_n\right)\right\|_{H_U}$$
$$=\left\|A^*F\left(\exp\left(i\arg AU_n\right)\right)-A^*AU_n-\alpha U_n\right\|_{H_U} \tag{109}$$

in the space H_U.

We denote the operator in right part of (106) as:

$$D(U) = \alpha^{-1}\left(E + \alpha^{-1}A^*A\right)^{-1}$$
$$\cdot A^*\left(F\exp\left(i\arg\left(AU\right)\right)\right).$$

$$(110)$$

For the operator $D(U)$ is valid Lemma 5.1. Let $A : H_U \to H_f$ be a completely continuous operator. Then the operator $D(U)$ defined by (110), is compact and it transfers any bounded set $U_r = \left\{U : \|U\|_{H_U} \le r\right\}$ into its relatively compact part at $\alpha^{-1}\|A^*\|\|F\|_{H_f} \le r < \infty$.

Thus, it is shown that there is true Corollary 5.2. If $\mathrm{grad}\,\sigma_a(U)$ is operator continuous in some neighborhood $U_* \subset H_U$ of the point U_*, then from Theorem 5.2 and Lemma 5.1 follows that the subsequence $\left\{U_{n_k}\right\}$ converges to some solution of (12) by the norm of the space H_U if $U_0 \in U_*$.

Dependent on the choice of initial approach the successive approximations (106) can converge to the solutions of various types [56-58].

Numerical Solution of Synthesis Problems with Use of the Energy Criterion σ_F

First we shall consider the iterative process of solution the equation of type (34) in the Hilbertian space $H_U = L_2(V) \otimes L_2(V) \otimes L_2(V)$ under certain restrictions on the parameter α . This equation is written as

$$\left(E - \frac{2}{\alpha}A^*N_0A\right)U = -\frac{2}{\alpha}A^*\left(|AU|^2 \cdot AU\right)$$

$$,(111)$$

where $E : H_U \to H_U$ is an identity operator,

$A : H_U \to C(\bar{G})$ is completely continuous operator. We denote $\|A\|_{H_U \to C(G)} = \beta$. Note that the scalar product and the corresponding norm $\|\cdot\|_{H_U}$ in the space H_U are defined by (14) and (15), and the Chebyshev $\|\cdot\|_{C(G)}$ and mean-square $\|\cdot\|_{C_{(G)}^{(2)}}$ norms in the space $C_{(G)}^{(2)}$ are introduced by Formulas (16)-(18).

Henceforth we shall consider completion of the space

$C_{(\bar{G})}^{(2)}$ relatively to the norm $\|\cdot\|_{C_{(\bar{G})}^{(2)}}$ [14], which is the Banach space and coincides with the Hilbertian space $H_f = L_2(\bar{G}) \otimes L_2(\bar{G})$, the norm in which we shall denote by symbol $\|\cdot\|_{l_2(G)}$. We assume that $A: H_U \to H_f$

is a completely continuous operator and in the space $C_{(\bar{G})}^{(2)}$ the domain of its values $R(A)$ is a set of continuous functions.

Taking into account the equality $AU = f$ we shall consider the expression $N_0 AU$ in (111) as an operator of multiplication by the function N_0:

$$N(f) = N_0 \cdot f,\qquad (112)$$

acting in the space H_f where N_0 is real nonnegative continuous function on the compact $\bar{G}$, in addition $\|N_0\|_{C(G)} = 1$. Obviously, that (112) is a linear bounded operator, and $\|N\|_{H_f \to H_f} \leq 1$.

If $\alpha > 2\|A^* NA\|$, then there exists the inverse operator

$\left(E - \frac{2}{\alpha} A^* NA\right)^{-1}$, the norm of which satisfies the inequality [14].

$$\left\|\left(E - \frac{2}{\alpha} A^* NA\right)^{-1}\right\| \leq \frac{1}{1 - \frac{2}{\alpha}\|A^* NA\|}.\qquad (113)$$

In this case, Equation (111) we shall write as

$$U = D(U)$$
$$\equiv -\frac{2}{\alpha}\left(E - \frac{2}{\alpha} A^* NU\right)^{-1} A^*\left(|AU|^2 \cdot AU\right).\qquad (114)$$

Here we shall show that the solution of (114) can be obtained as a limit of successive approximations of the iterative process [61]:

$$U_{n+1} = tU_n + (1-t)D(U_n) \quad (n = 0,1,2,\cdots),$$

(115)

where t is some fixed number with the interval $(0,1)$. In addition successive approximations can converges to different solutions of (114) depending on the choice of the initial approximation.

To determine the conditions and to justify convergence of (115), we shall use the Theorem 4.1 with [61, p. 68], according to which: if nonexpanding operator W converts a closed convex set ω of strictly convex Banach space X into its compact part, then successive approximations

$$x_{n+1} = tx_n + (1-t)W(x_n) \quad (n = 0,1,2,\cdots)$$ where t is any fixed number from the interval $(0,1)$, converges to some solution of the equation $x = U(x)$ at some $x_0 \in \omega$.

Since the Hilbertian space H_U is strictly convex Banach space (see [61, p. 67]), then to satisfy of the conditions of this theorem concerning (114), it is sufficiently to show that a closed convex set S_{r_0} exists in the space H_U, where the operator $D(U)$ is nonexpanding and completely continuous. In addition there is such relation

$$D\left(S_{r_0}\right) \subset S_{r_0}.$$

Satisfication of these conditions results from lemmas, proved in [7,49].

Lemma 5.2. Let $A: H_U \to H_f$ be a linear completely continuous operator and the domain of its values $R(A)$ is a set of continuous functions, $\alpha > 2\|A^* N_0 A\|$. Then $D(U)$ is a nonexpanding operator on

$$S_{r_0} \subset L_2(\bar{V}),$$

where

$$S_{r_0} = \left\{ U : \|U\|_{H_U} \le r_0 \right\},$$

$$r_0 = \left(\frac{1 - \mu\|A^* N_0 A\|}{3\mu\beta^3 \|A^*\|} \right)^{1/2}, \mu = 2/\alpha.$$

(116)

that is, for any $U_1, U_2 \in S_{r_0}$ the inequality

$$\|D(U_1) - D(U_2)\|_{H_U} \le \|U_1 - U_2\|_{H_U}$$

(117)

is satisfied.

Lemma 5.3. Let $A: H_U \to H_f$ be a linear completely continuous operator and the domain of its values $R(A)$ is a set of continuous functions, $\alpha > 2\|A^* NA\|$. Then $D(U): H_U \to H_U$, defined by (114), is a completely continuous operator for which the relation

$$D\left(S_{r_0}\right) \subset S_{r_0} \quad (S_{r_0} \text{ is a closed convex set, defined by}$$

(116)), is satisfied.

Numerical Solution of Synthesis Problems with Optimization of Geometry of Radiating System

In this section we shall consider the synthesis problem of a flat aperture according to the prescribed amplitude DP for the case when the form of aperture and amplitudephase distribution of the field (currents) in it is optimized simultaneously, limiting by the case of linear polarization [25,26,77]. We shall consider a special case when the field in the aperture is linearly polarized along one of the coordinate axes, and DP has only one component. We introduce inside of aperture the polar coordinate system: $x = r\cos\psi$, $y = r\sin\psi$. Let $\rho(\psi)$ be a function of the boundary of aperture $\bar{S}$. Then DP $f(s_1,s_2)$ which is formed by amplitude-phase distribution of the field in the aperture $U(r,\psi)$, is given by the formula [7,16]

$$f_{g_j}(s_1,s_2) = A_j(U_v,\rho)$$

$$\equiv \int_0^{2\pi} \int_0^{\rho(\psi)} U_v(r,\psi) e^{ikr(s_1\cos\psi + s_2\sin\psi)} r\, dr\, d\psi$$

$$(j=1, v=x;\ \ j=2, v=y).$$
(118)

Later on we omit the index in definition of f. Let the given amplitude DP $F(s_1,s_2)$ be different from identical zero in some limited closed domain $\bar{G} \subset \Omega = \mathbb{R}^2$ and it is identically equal to zero at $(s_1,s_2) \in \Omega \setminus \bar{G}$. The problem of simultaneous synthesis of the aperture shape S and amplitude-phase distribution of the field in it is considered as the problem on finding the functions $U(r,\psi)$ and $\rho(\psi)$ minimizing the functional

$$\tilde{\sigma}_F(U,\rho) = \iint_G \left[F(s_1,s_2) - |f(s_1,s_2)| \right]^2 ds_1 ds_2$$

$$+ \iint_{\Omega \setminus G} |f(s_1,s_2)|^2 ds_1 ds_2 + \gamma \int_0^{2\pi} \int_0^{\rho(\psi)} r\, dr\, d\psi,$$
(119)

in which the first two summands describe the meansquare deviation of modules of given and synthesized DP's in space $\mathbb{R}^2$, and the third one—imposes restrictions on the square of aperture S. We shall consider the parameter $\gamma > 0$ as a weight coefficient. We introduce into consideration the following functional spaces: $H_U = L_2(S)$ is a space of square integrable complex functions in the domain S, $H_\rho = L_2[0, 2\pi]$ is a space of square integrable real functions on the segment $[0, 2\pi]$, $H_f = L_2(\Omega)$ is a space of square integrable complex functions in the domain Ω. Scalar products and generated by it norms we shall introduce as follows:

$$(U_1, U_2)_{H_U} = \int_0^{2\pi} \int_0^{\rho(\psi)} U_1(r, \psi) \overline{U_2(r, \psi)} r dr d\psi$$

$$\|U\|_{H_U} = (U, U)^{1/2}$$

$$(\rho_1, \rho_2)_{H_\rho} = (1/2) \int_0^{2\pi} \rho_1(\psi) \rho_2(\psi) d\psi$$

$$\|\rho\|_{H_\rho} = (\rho, \rho)^{1/2}$$

$$(f_1, f_2)_{H_f} = \iint_\Omega f_1(s_1, s_2) \overline{f_2(s_1, s_2)} ds_1 ds_2$$

$$\|f\|_{H_f} = (f_1, f_2)^{1/2}.$$ (120)

Taking into account the introduced norms, the last summand in (119) and Parseval's equality have the form

$$\gamma \int_0^{2\pi} \int_0^{\rho(\psi)} r dr d\psi = \frac{\gamma}{2} \int_0^{2\pi} \rho^2(\psi) d\psi = \gamma \|\rho\|_{H_\rho}^2,$$

$$\|f\|_{H_f}^2 = (2\pi/k)^2 \|U\|_{H_U}^2.$$

On this base the functional $\tilde{\sigma}_F$ is presented as:

$$\tilde{\sigma}_F(U, \rho) = \|F\|_{H_f}^2 - 2(F, |f|)$$
$$+ (2\pi/k)^2 \|U\|_{H_U}^2 + \gamma \|\rho\|_{H_\rho}^2.$$ (121)

We shall consider the iterative process of numerical minimization of (121). In it base we shall put the ideas similar, as at minimization of functions of two variables by a coordinate descent method. Let $V^* = (U^*, \rho^*)$ be a minimum point of the functional $\tilde{\sigma}_F(U, \rho)$ and

$V^{(0)} = (U^{(0)}, \rho^{(0)})$ be an initial approximation chosen from some neighborhood of the point V^*. We shall denote by $S^{(0)}$ the initial shape of aperture, that is described by the function $\rho^{(0)}(\psi)$. Substitute $\rho^{(0)}(\psi)$ in (121) and consider its restriction in the space H_U:

$$\tilde{\sigma}_U(U) = \tilde{\sigma}_F\left(U, \rho^{(0)}\right).$$

(122)

From the necessary condition of the functional minimum $\tilde{\sigma}_U(U)$ we obtain equation of type (9). Numerically we solve it by successive approximations method, given in pt. 5.1:

$$f_{n+1}(Q') = (k/2\pi)^2 \iint_{\bar{G}} F(Q) K(Q',Q;k) e^{i \arg f_n(Q)} dQ$$

$$(n = 0,1,2,\cdots).$$

(123)

As a result, we find the function $f^{(1)}(Q)$, and obtain the first approximation of the solution $U^{(1)}$ by the formula of type (10).

We shall pass to finding the function $\rho(\psi)$ that describes the boundary of aperture $\bar{S}$. We fix the function $U^{(1)}$ extending its analytically according to (10) to the plane XOY in (121), and consider the functional

$$\tilde{\sigma}_\rho(\rho) = \tilde{\sigma}_F\left(U^{(1)}, \rho\right)$$ which depends only on the function ρ. With the necessary minimum condition:

$$\left(\operatorname{grad}\tilde{\sigma}_\rho(\rho), g\right) = \left(\tilde{\sigma}'_\rho(\rho), g\right) = 0$$, where

$$\left(\tilde{\sigma}'_\rho(\rho), g\right) = 0$$ is an arbitrary element of the space H_ρ, we obtain the equation

$$B(\rho)$$

$$= \rho(\psi) \left\{ \left| \left(\frac{k}{2\pi}\right)^2 \iint_{\bar{G}} F(Q) e^{i \arg f(Q)} e^{-ik\rho(\psi)\alpha(Q,\psi)} dQ \right|^2 \right.$$

$$\left. -\gamma\left(\frac{k}{2\pi}\right)^2 \right\} = 0$$

(124)

which is a nonlinear functional equation with respect to the function $\rho(\psi)$.

We shall find numerically solutions of (124), using the Newton-Kantorovich method [69]:

$$B'(\rho_n)\Delta\rho_n = -B(\rho_n),$$

(125)

$$\rho_{n+1}(\psi) = \rho_n(\psi) + \Delta\rho_n(\psi)$$
$$(n = 0, 1, 2, \cdots)$$

,

(126)

where $B'(\rho_n)$ is the partial Frechet derivates of operator B by the function ρ. We assume that $\rho_0 = \rho^{(0)}$. Equation (125) is a linear integral equation of the form

$$L(\rho_n(\psi))\Delta\rho_n(\psi)$$
$$+ \int_0^{2\pi} M[\rho_n(\psi), \rho_n(\psi'), \psi, \psi']\Delta\rho_n(\psi')d\psi'$$
$$= -B(\rho_n(\psi))$$

(127)

which $L(\rho_n(\psi)) \neq 0$ can be reduced to the Fredholm equation of the second kind at $L(\rho_n(\psi)) \neq 0$. Solving (127) we find the first approximation for the function $\rho^{(1)}$ that describes a boundary of aperture of the radiating system.

Continuing finding in turn the approximations of functions $U^{(n)}$ and $\rho^{(n)}$, we obtain the sequence $\{U^{(n)}, \rho^{(n)}\}$ that is relaxational for (119). In more detail the problem of choice of initial approximations and justification of relaxation for functional $\tilde{\sigma}_F(U, \rho)$ is given in [7,56,57].

First we shall consider the numerical results of synthesis of flat aperture with optimization of its geometry. In Figures 15 and 16 the examples of synthesis of amplitude DPs, which in cross section have quasi-square and quasi-triangular shapes, are given. The optimal shapes of apertures are given there too.

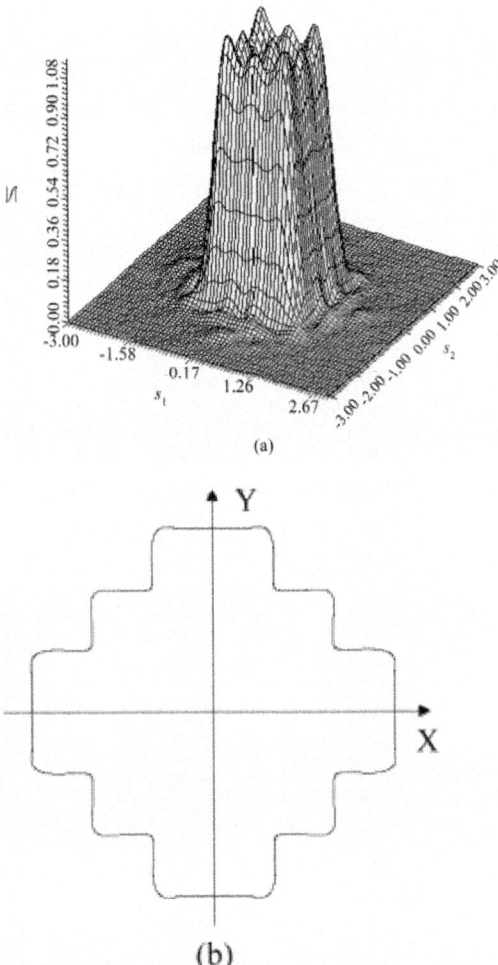

(a)

(b)

Figure 15: Synthesis DP with square contour: (a) Synthesized DP with square contour; (b) The optimal shape of aperture.

Note that the problems of such class arise, in particular, at the synthesis of contour DPs of fixed and variable forms for satellite antenna systems needed for uniform irradiation of a given territorial zone from the board of artificial satellite, where multi-beam antennas are used often [77- 82].

If multibeam antenna (MBA) has a radiating aperture of circular shape, and partial beams in the cross section have the shape of a circle and nonuniform distribution of radiated energy inside of the section, then on the junction of three neighboring rays with a circular cross section the so-called critical zones

(Figure 17) with low level of radiated energy occur. One of the possible ways of solution of this problem is passage to alternative forms of apertures that on the base of the optimal APD will form rays that have rectangular, triangular or hexagonal shapes and close to constant (inside of contour) coefficient of directed action in rectangular cross section.

Obviously that on the base of such partial beams it is easy to synthesize given summary DP without critical zones.

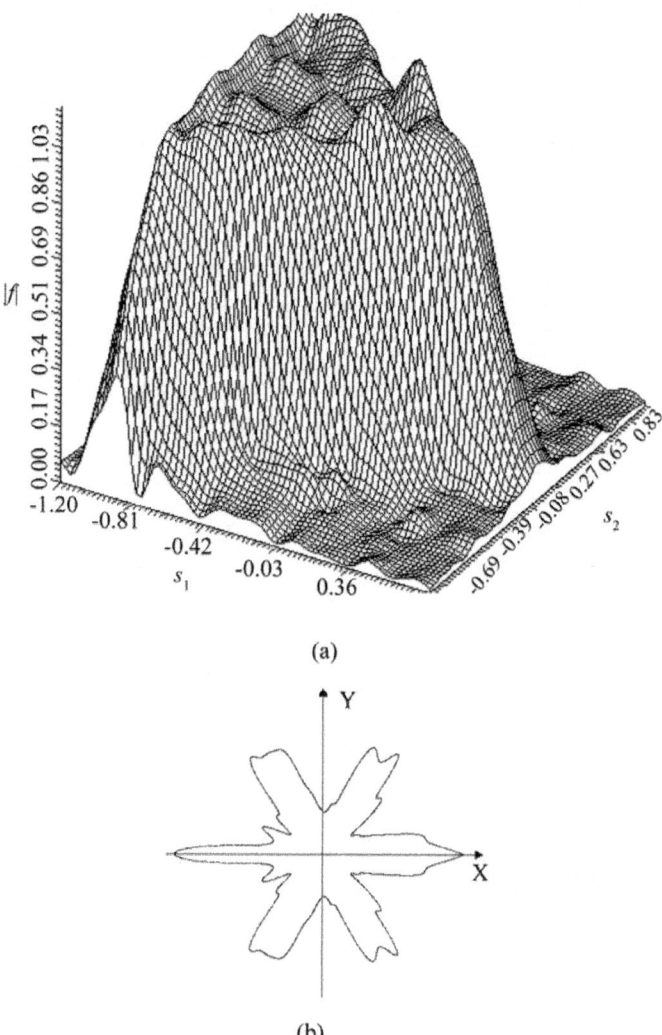

(a)

(b)

Figure 16: Synthesis of DP with triangular contour: (a) Synthesized DP with quasi-triangular contour; (b) Optimal shape of aperture.

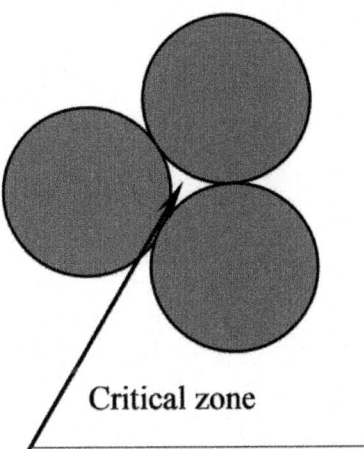

Figure 17: Critical zones with low level of radiated energy.

Below the results of synthesis of triangle-beam contour DP with partial beams with circular (Figure 18(a)) and quasi-rectangular (Figure 18(b)) contours are presented.

From the analysis of the figures we see that in the summary DP which is obtained on the base of quasisquare of contours, critical zones are absent, and variation of radiated energy inside of the contour does not exceed 2 dB.

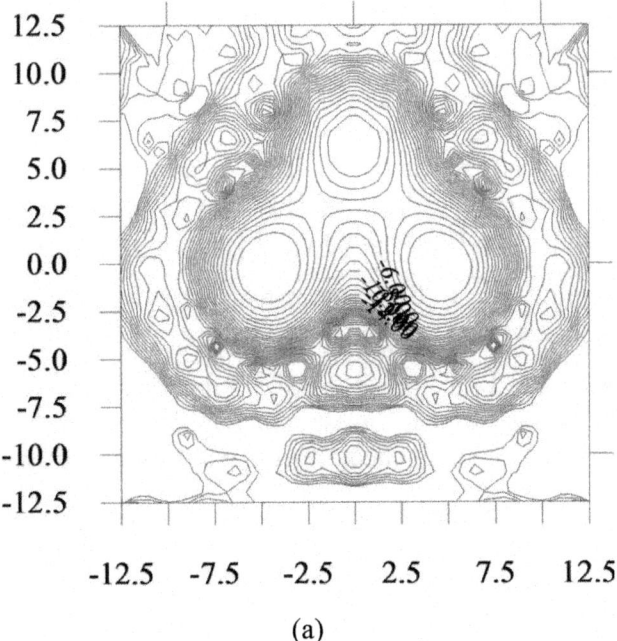

(a)

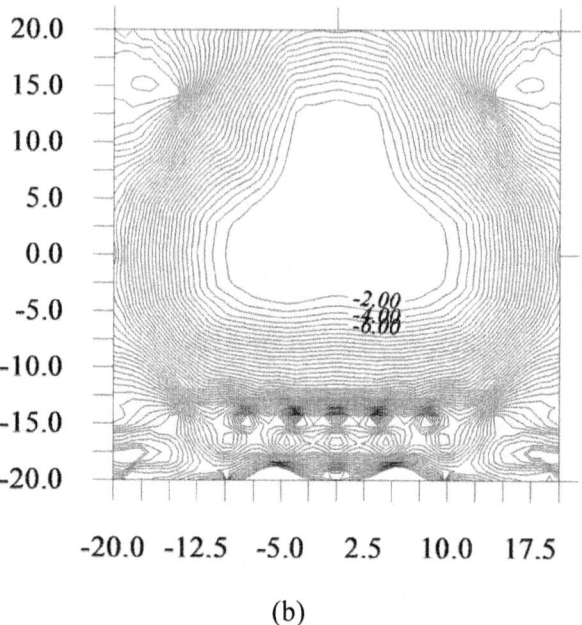

(b)

Figure 18: Synthesis of contour DP.

CONCLUSIONS

Note the main features and problems arising at investigations of othe class of problems reviewed in the article:

- The investigation of nonuniqueness and branching of existing solutions, which depend on the physical parameters of the problem is the main difficulty at solving this class of problems. As follows from the researches presented, in particular, for a special case in [7,13,16], when $F(s_1,s_2) = F_1(s_1) \cdot F_2(s_2)$, the quality of existing solutions increases significantly with growth of parameters c_1, c_2. However, to obtain the best approximation to the given amplitude DP $F(s_1,s_2)$ at relatively small values of the parameters c_1, c_2 describing the sizes of aperture, allowing to confine by investigations of the first few branching points (lines), is essential in the synthesis problems of radiating systems.

- At finding the solutions of (45) by the successive approximations method in the case of the even by both arguments (or one argument) functions $F(s_1,s_2)$ to obtain solution of certain type it is necessary to choose an initial approximation, which belongs to the corresponding invariant set of nonlinear operators B_1 and B_2.

- On the base of computational experiments it is revealed that the branching-off complex solutions, which exist at small sizes of aperture, increase the efficiency of synthesis within 20% - 40% compared with the real (primary) solutions. The presence of different by structure but identical by efficiency solutions (in the sense of value of the corresponding functional), provides for practice the possibility of choosing one of them that has a simpler physical realization. In addition, the branching-off solutions at conservation of the same efficiency which corresponds to real solutions allow to reduce the linear size of the aperture in the limits of 10 to 20 percent.

- The proposed numerical method of solution of nonlinear two-parameter spectral problems arising at investigation of nonlinear integral equations can be successfully applied, in particular, to solving the linear and nonlinear two-parameter spectral problems concerning matrix equations and ordinary differential equations of the second n-th order with nonlinear occurrence of the spectral parameters into coefficients of equations and boundary conditions.

- A mathematical analogy between the synthesis problems of acoustic and electromagnetic antennas and synthesis problems of radio allows to use developed methods and numerical algorithms in the above sections of acoustics, radio physics and radio engineering.

- In mathematical aspect the synthesis problems of radiating systems formulated in paragraph 2, belong to problems of non-linear approximation of real finite functions by modules of one-dimensional or two-dimensional, or else discrete Fourier transforms [56,57]. In this connection, the above results can be directly applicable to solving the problems of mean-square approximation of nonnegative finite functions.

REFERENCES

1. G. T. Markov, B. M. Petrov and G. P. Grudinskaja, "Electrodynamics and Propagation of Radio Waves," Sov. Radio, Moscow, 1979.

2. G. T. Markov and A. F. Chaplin, "Excitation of Electromagnetic Waves," Radio and Communications, Moscow, 1983.

3. V. B. Zhukov, "Calculation of the Hydroacoustic Antennas According to

the Directivity Pattern," Sudostroienije, Leningrad, 1977.

4. A. F. Chaplin, "Analysis and Synthesis of Antenna Array," Vyschaya Shkola, Lvov, 1987.

5. G. T. Markov and D. N. Sazonov, "Antennas," Energiya, Moscow, 1975.

6. E. G. Zelkin and V. G. Sokolov, "Synthesis Methods of Antenna Arrays and Antenna with Continuous Aperture," Sov. Radio, Moscow, 1980.

7. P. O. Savenko, "Nonlinear Problems of Radiating Systems Synthesis (Theory and Methods of Solution)," IAPMM of NASU, Lviv, 2002.

8. B. M. Minkovich and V. P. Yakovlev, "Theory of Synthesis of Antennas," Sov. Radio, Moscow, 1969.

9. B. Z. Katsenelenbaum, "Approximability Problem of the Electromagnetic Field," Successes of Physics Sciences, Vol. 164, No. 9, 1994, pp. 983-993.doi:10.3367/UFNr.0164.199409f.0983

10. G. V. Aleksejev, "About Incorrect of the First Kind Nonlinear Operator Equation of Antennas Synthesis Theory," Journal of Computational Mathematics and Mathematical Physics, Vol. 19, No. 6, 1979, pp. 1590-1595.

11. T. N. Tikhonov and V. Y. Arsenin, "The Methods of Solution of Incorrect Problems," Nauka, Moscow, 1979.

12. V. K. Ivanov, V. V. Vasin and V. P. Tanana, "The Theory of Linear Incorret Problems and its Applications," Nauka, Moscow, 1978.

13. M. I. Andriychuk, N. N. Voitovich, P. A. Savenko and V. P. Tkachuk, "Antenna Synthesis According to Amplitude Directivity Pattern: Numerical Methods and Algorithms," Naukova Dumka, Kiev, 1993.

14. V. A. Trenogin, "Functional Analysis," Nauka, Moscow, 1980.

15. N. N. Voitovich and P. A. Savenko, "Generalized Criterion of Proximity Diagram in the Antennas Synthesis Problem by Method V. V. Semjonova," Radio Engineering and Electronics, Vol. 17, No. 9, 1973, pp. 1794-1798.

16. N. N. Voitovich and P. A. Savenko, "Branching of Solutions of the Antenna Synthesis Problem According to the Prescribed Amplitude Directivity Pattern," Radio Engineering and Electronics, Vol. 21, No. 4, 1976, pp. 723- 729.

17. N. N. Voitovich and P. A. Savenko, "Synthesis of Antennas According to the Prescribed Amplitude Directivity Pattern and Associated Problems of Quasi-optics (Review)," Radio Engineering and Electronics, Vol. 24, No. 8, 1979, pp. 1465-1500.

18. P. A. Savenko, "Numerical Solution of One Class of Nonlinear Problems of Synthesis Theory of Radiating Systems," Journal of Computational

Mathematics and Mathematical Physics, Vol. 40, No. 6, 2000, pp. 929-939.

19. P. O. Savenko, "About the Existence of Solutions of One Class of Nonlinear Inverse Problems of Mathematical Physics Concerning Synthesis of Radiating Systems," Reports of Academy of Sciences of Ukraine, No. 8, 1994, pp. 48-52.

20. A. N. Kolmogorov and S. V. Fomin, "Elements of the Theory of Functions and Functional Analysis," Nauka, Moscow, 1968.

21. M. M. Vainberg, "Variational Method and Method of Monotone Operators in the Theory of Nonlinear Equations," Nauka, Moscow, 1972.

22. P. A. Savenko, "Branching of Solutions of the Antenna Synthesis Problem According to the Prescribed Amplitude Directivity Pattern with Use of Regularizing Functionals," Proceedings of the Institute of Higher Learning, Radio Electronics, Vol. 39, No. 2, 1996, pp. 35-50.

23. A. F. Verlan and V. S. Sizikov, "Integral Equations: Methods, Algorithms, Programs. Reference Textbook," Naukova Dumka, Kyiv, 1986.

24. P. O. Savenko, "Synthesis of Linear Antenna and Antenna Arrays with Use of Regularizing Functionals," In: Proceedings of Interantional Seminar/Workshop "Direct and Inverse Problems of Electromagnetic and Acoustic Wave Theory", IAPMM of NASU, Lviv, 1995, pp. 62-63.

25. P. O. Savenko and V. J. Anokhin, "Numerical Optimization of Shape of Flat Aperture and Field According to the Prescribed Amplitude Directivity Pattern," Theoretical Electrical Engineering, Vol. 53, 1996, pp. 118-126.

26. P. O. Savenko and V. J. Anokhin, "Synthesis of the Surface Antenna for a Given Power Pattern," VI-th International Conference "Mathematical Methods in Electromagnetic Theory", Lviv, 10-13 September 1996, pp. 525-528.

27. V. I. Dmitriev and N. I. Berezina, "Numerical Methods for Solving the synthesis of Radiating Systems," MGU, Moscow, 1986.

28. P. A. Savenko, "Synthesis of Linear Antenna Arrays According to the Prescribed Amplitude Directivity Pattern," Proceedings of the Institute of Higher Learning, Radio Physics, Vol. 22, No. 12, 1979, pp. 1498-1504.

29. P. A. Savenko and L. M. Pasnak, "Numerically-Analytical Method of Synthesis of Linear Antenna Arrays of Vibrators According to the Prescribed Amplitude Directivity Pattern," Proceedings of the Institute of Higher Learning, Radio Physics, Vol. 40, No. 12, 1997, pp. 11-25.

30. P. A. Savenko and L. M. Pasnak, "The Numerical Solution of Nonlinear Synthesis Problem of Antenna Array of Curved Radiator with Account of

their Mutual Influence," Proceedings of the Institute of Higher Learning, Radio Electronics, Vol. 43, No. 8, 2000, pp. 1-15.

31. P. A. Savenko and M. D. Tkach, "Synthesis of Flat Microstrip Antenna Array According to the Prescribed Energy Directivity Pattern," Proceedings of the Institute of Higher Learning, Radio Electronics, Vol. 44, No. 6, 2001, pp. 19-32.

32. L. M. Pasnak and P. O. Savenko, "Nonlinear Synthesis Problems of Antenna Arrays Considering Mutual Influence of Scatterers," Millennium Conf. on Antennas & Propagation, Davos, 9-14 April 2000, p. 201.

33. L. M. Pasnak and P. O. Savenko, "Synthesis of Antenna Arrays with Regard to the Radiators Mutual Coupling and Vector Characteristics of Electromagnetic Fields," III International Conference on Antenna Theory and Techniques (ICATT'99), Sevastopol, 8-11 September 1999, pp. 135-137.

34. L. M. Pasnak and P. O. Savenko, "Numerical Solution of Phase Synthesis Problem of Linear Array of Vibrators with Account their Mutual Influence," Proceedings of the Institute of Higher Learning, Radio Electronics, Vol. 40, No. 9, 1997, pp. 50-62.

35. V. F. Kravchenko, L. P. Protsakh, P. A. Savenko and M. D. Tkach, "The Mathematical Features of the Synthesis of Plane Equidistant Antenna Arrays According to the Prescribed Amplitude Directivity Pattern," Antennas, Vol. 154, No. 3, 2010, pp. 34-48.

36. P. A. Savenko and M. D. Tkach, "The Numerical Solution of the Nonlinear Synthesis Problem of Microstrip Antenna Array with Account the Mutual Influence of Radiators," Radio Engineering and Electronics, Vol. 46, No. 1, 2001, pp. 58-65.

37. P. O. Savenko, "Structure of Solutions of Synthesis Problem of Microstrip Array According to the Prescribed Energy Directivity Pattern," Mathematical Methods and Physicomechanical Fields, Vol. 46, No. 3, 2003, pp. 151- 163.

38. L. D. Bakhakh and S. D. Kremenetskii, "Synthesis of Radiating Systems (Theory and Methods of Calculations)," Sov. Radio, Moscow, 1974.

39. V. I. Dmitriev, A. S. Il'inskii and A. G. Sveshnikov, "Development of Mathematical Methods for the Study of Direct and Inverse Problems of Electrodynamics," Successes of Mathematical Sciences, Vol. 31, No. 6, 1976, pp. 123-141.

40. V. I. Dmitriev and P. P. Sereda, "Mathematical Models and Method of Integral Equations in the Theory of Spiral Wire Antennas," Numerical Methods of Electrodynamics, Vol. 4, 1980, pp. 55-65.

41. A. S. Il'inskii, "Electrodynamic Theory of Antenna Arrays," Proceedings of the Institute of Higher Learning, Radio Electronics, Vol. 21, No. 2, 1978, pp. 9-21.

42. A. S. Il'inskii and I. V. Berezhnaja, "Investigation of the Current Distribution in the System Arbitrarily Located Vibrators," Computational Methods and Programming, Vol. 20, 1973, pp. 263-268.

43. L. M. Pasnak and P. O. Savenko, "Synthesis of Linear Equidistant Antenna Array of Interacting Vibrators According to the Prescribed Amplitude Directivity Pattern," Radio Engineering and Electronics, Vol. 42, No. 5, 1997, pp. 567- 572.

44. P. O. Savenko and L. M. Pasnak, "Synthesis of Linear Antenna Array for According to the Prescribed Amplitude DP with Account of their Mutual Influence," Bulletin of Lviv University, Vol. 42, 1995, pp. 69-74.

45. P. A. Savenko and L. M. Pasnak, "Numerical Solution of the Nonlinear Synthesis Problem of Antenna Array of Curved Radiators with Account their Mutual influence," Proceedings of the Institute of Higher Learning, Radio Electronics, Vol. 43, No. 8, 2000, pp. 1-15.

46. P. O. Savenko "About One Method of the Synthesis Problems of Radiating Systems under the Given Power Directivity Pattern," International Conference of "Mathematic Methods in Electromagnetic Theory", Kharkiv, 11-15 September 2000, pp. 618-620.

47. P. A. Savenko, "About Structure of the Solutions of the Synthesis of Linear Antenna under the Given Energy Directivity Pattern," Radio Physics and Radio Astronomy, Vol. 5, No. 4, 2000, pp. 405-415.

48. P. O. Savenko, "Numerical Solution of Synthesis Problems of Radiating Systems According to the Prescribed Power Directivity Pattern," Proceeding of 6th International Seminar "Direct and Inverse Problem of Electromagnetic and Acoustic Wave Theory", IAPMM of NASU, Lviv, 2001, pp. 141-145.

49. P. A. Savenko, "The Numerical Solution of the Inverse Synthesis Problems of Radiating Systems According to the Prescribed Power Directivity Pattern," Journal of Computational Mathematics and Mathematical Physics, Vol. 42, No. 10, 2002, pp. 1495-1509.

50. M. M. Vainberg, "Variational Methods of Investigation of Nonlinear Operators," Publishing House of TechnicoTheoretical Literature, Moscow, 1956.

51. M. M. Vainberg, "Functional Analysis," Prosveshchenije, Moscow, 1979.

52. P. O. Savenko, "Branching of Solutions of the Synthesis Problem of Linear

Antenna According to the Prescribed Amplitude Directivity Pattern in the Space of Helder," Mathematical Methods and Physicomechanical Fields, Vol. 41, No. 2, 1998, pp. 35-44.

53. M. M. Vainberg and V. A. Trenogin, "Theory of Branching of Solutions to Nonlinear Equations," Nauka, Moscow, 1969.

54. N. N. Voitovich and P. A. Savenko, "About One Integral Equation of Antenna Synthesis Theory," Mathematical Methods and Physicomechanical Fields, 1975, pp. 161- 163.

55. P. A. Savenko, "About the Branching of the Solutions of the Synthesis Problems of Radiating Systems with Flat Aperture According to the Prescribed Amplitude Directivity Pattern," Izv. Vysch. Uch. Zaved., Radiophys, Vol. 29, No. 3, 1986, pp. 339-345.

56. P. Savenko and M. Tkach, "Numerical Solution of MeanSquare Approximation Problem of Real Nonnegative Function by the Modulus of Double Fourier Integral," Applied Mathematics, Vol. 2, No. 2, 2011, pp. 1076-1090. doi:10.4236/am.2011.29149

57. P. Savenko and M. Tkach, "Numerical Approximation of Real Finite Nonnegative Function by the Modulus of Discrete Fourier Transformation," Applied Mathematics, Vol. 1, No. 1, 2010, pp. 65-75. doi:10.4236/am.2010.11008

58. L. P. Protsakh, P. O. Savenko and M. D. Tkach, "Branching of Solutions of the Problem of Mean-square Approximation of Real Finite Function with Respect to Two Variables by Modulus of Double Fourier Transformation," Mathematical Methods and Physicomechanical Fields, Vol. 53, No. 3, 2010, pp. 3-7.

59. I. I. Liashko, V. F. Yemelianov and A. K. Boyarchuk, "Bases of Classical and Modern Mathematical Analysis," Vyshcha Shkola, Kyiv, 1988.

60. I. I. Privalov, "Introduction to the Theory of Functions of Complex Variables," Nauka, Moscow, 1984.

61. M. A. Krasnosel'skii, G. M. Vainikko and P. P. Zabreiko, "Approximate Solution of Operational Equations," Nauka, Moscow, 1969.

62. A. A. Abramov and L. F. Yukhno, "About a General Nonlinear Self-Adjoint Spectral Problem for Ordinary Differential Equations Systems," Journal of Computational Mathematics and Mathematical Physics, Vol. 49, No. 4, 2009, pp. 624-627.

63. S. I. Solov'yev, "The Finite Element Method for Symmetric Eigenvalue Problems with Nonlinear Occurrence of Spectral Parameter," Journal of Computational Mathematics and Mathematical Physics, Vol. 37, No. 11,

1997, pp. 1311-1318.

64. L. P. Protsakh, P. O. Savenko and M. D. Tkach, "Numerical Solving of Two-Dimensional Nonlinear Spectral Problem Arising in Antenna Synthesis with Flat Radiating Aperture," Proceedings of Xth International Seminar /Workshop "Direct and Inverse Problems of Electromagnetic and Acoustic Wave Theory", Lviv, 23-25 September 2005, pp. 199-203.

65. P. A. Savenko and L. P. Protsakh, "Implicit Function Method in Solving the Two-Dimensional Nonlinear Spectral Problem," Proceedings of the Institute of Higher Learning, Mathematics, Vol. 546, No. 3, 2007, pp. 41-44.

66. L. P. Protsakh and P. O. Savenko, "Implicit Functions Methods in Solving the Two-Parameter Linear Spectral Problems," Mathematical Methods and Physicomechanical Fields, Vol. 52, No. 2, 2009, pp. 42-49.

67. P. O. Savenko and L. P. Protsakh, "Numerical Solution of Two-Point Boundary Problem with Two-Dimensional Nonlinear Spectral Parameter," Mathematical Methods and Physicomechanical Fields, Vol. 1, No. 1, 2011, pp. 48-56.

68. G. M. Vainikko, "Analysis of Discretized Methods," Tartu State University, Tartu, 1976.

69. L. V. Kantorovich and G. P. Akilov, "Functional Analysis," Nauka, Moscow, 1977.

70. V. I. Smirnov, "Course of High Mathematics," Nauka, Moscow, 1974.

71. P. O. Savenko, "The Numerical Algorithm of Solving the Generalized Eigenvalues Problem for Completely Continuous Self-Adjoint Operators with Nonlinear Spectral Parameter," Mathematical Methods and Physicomechanical Fields, Vol. 40, No. 1, 1997, pp. 146-150.

72. [73] P. O. Savenko and L. P. Protsakh, "Variational Approach to Solving the Nonlinear Vector Spectral Problem for the Case of Self-Adjoint Positive Semi-definite Operators," Reports of National Academy of Sciences of Ukraine, No. 6, 2004, pp. 26-31.

73. P. A. Savenko, "The Branching of the Solutions of the Synthesis Problems of Radiating Systems with Flat Aperture According to the Prescribed Amplitude Directivity Pattern," Proceedings of the Institute of Higher Learning, Radio Physics, Vol. 29, No. 3, 1986, pp. 339-345.

74. L. P. Protsakh, P. O. Savenko and M. D. Tkach, "Branching of Solutions for the Problem of Meansquare Approximation of a Real Finite Function of Two Variables by the Modulus of Double Fourier Transformation," Journal of Mathematical Sciences, Vol. 180, No. 1, 2012, pp. 51- 67.

75. P. A. Savenko, "The Numerical Solution of One Class of Nonlinear Synthesis Problems of Radiating Systems," Journal of Computational Mathematics and Mathematical Physics, Vol. 40, No. 6, 2000, pp. 929-939.

76. P. O. Savenko, "About Convergence of the Iterative Process in a Nonlinear Synthesis Problems of Radiating Systems under the Use Smoothing Functionals," Mathematical Methods and Physicomechanical Fields, Vol. 43, No. 2, 2000, pp. 105-111.

77. P. O. Savenko and V. J. Anokhin, "Synthesis of Amplitude-Phase Distribution and Shape of a Plane Antenna Aperture for a Given Power Pattern," IEEE Transactions on Antennas and Propagation, Vol. 45, No. 4, 1997, pp. 744-747. doi:10.1109/8.564102

78. V. J. Anokhin and P. O. Savenko, "Synthesis of the Contour Directivity Pattern without Critical Zones," URSI International Symposium on Electromagnetic Theory, St. Peterburg, 23-26 May 1995, pp. 197-199.

79. P. O. Savenko and V. J. Anokhin, "Synthesis of the Surface Antenna for a Given Power Pattern," 6th International Conference "Mathematical Methods in Electromagnetic Theory", Lviv, 10-13 September 1996, pp. 525-528.

80. B. Podlevskyi and P. Savenko, "Synthesis of the Multibeams Reflector Antenna Directivity Patterns on the Given Magnitude," 6th International Conference "Mathematical Methods in Electromagnetic Theory", Lviv, 10- 13 September 1996, pp. 521-524.

81. B. Podlevskyi and P. Savenko, "Synthesis of the Variable Shape Contoured Directivity Patterns of the Hybrid Reflector Antennas," Transactions on Black Sea Region Symposium on Applied Electromagnetism, Athens, 17-19 April 1996, p. 20.

82. B. Podlevskyi and P. Savenko, "Synthesis of the Variable Shape Contoured Directivity Patterns of the Hybrid Reflector Antennas," Electromagnetics, Vol. 18, No. 5, 1998, pp. 507-527. doi:10.1080/02726349808908607

Chapter 4

CONTINUOUS HYBRID MULTISTEP METHODS WITH LEGENDRE BASIS FUNCTION FOR DIRECT TREATMENT OF SECOND ORDER STIFF ODES

Olabode B. T., Momoh A. L.

Department of Mathematical Sciences, Federal University of Technology, Akure, Nigeria

ABSTRACT

This article proposed continuous hybrid multistep methods with Legendre polynomial as basis functions for the direct solution of system of second order ordinary differential equations. This was achieved by constructing a continuous representation of hybrid multistep schemes via interpolation of the approximate solution and collocation of derivative function with Legendre polynomial as basis functions. The discrete schemes were obtained from the continuous scheme as a by-product and applied in block form as simultaneous numerical integrators to solve initial value problems (IVPs). The resultant schemes are self-starting, do not need the development of separate predictors, consistent, zero-stable and convergent. The performance of the methods was demonstrated on some numerical examples to show accuracy and efficiency advantages. The numerical results compared favourably with existing method.

INTRODUCTION

The mathematical modeling of physical phenomena in science and engineering field especially in mechanical systems with several springs attached in series or dissipation, control theory, celestial mechanics, series circuits lead to a system of differential equations (see Landau and Lifshitz (1965), Liboff (1980)). Realistically, the analytical solutions of most differential equations are not easily obtainable. This necessitated the need for approximate solution by the application of numerical techniques.

The techniques for the derivation of continuous linear multistep methods (LMMs) for direct solution of initial value problems in ordinary differential equations have been discussed in literature over the years and these include, among others collocation, interpolation, integration and interpolation polynomials. Basis functions such as, power series, Chebyshev polynomials, trigonometric functions, the monomials x^r, the canonical polynomial $(Q_r(x), r \geq 0)$ of the Lanczos Tau method in a perturbed collocation approach have been employed for this purpose (see Abualnaja (2015); Adeyefa *et al.*, (2014); Awoyemi and Idowu (2005); Lambert (1991)).

Moreover, power series has also being extensively used in literature for the same purpose. Sirisena *et al.*, (2004) proposed a new Butcher type two-step block hybrid multistep method for accurate and efficient parallel solution of order ordinary differential equations. Awoyemi and Idowu (2005) developed a class of hybrid collocation methods for third order ordinary differential equations with power series as the basis functions and were implemented in predictor–corrector mode. Ehigie *et al.*, (2010) worked on generalized two-step continuous linear multistep method of hybrid type for the direct integration of second order ordinary differential equations. Fudzial *et al.*, (2009) constructed the explicit and implicit 3-point-1- block (I3P1B) for solving special second order ordinary equations directly. Awari (2013) considered the derivation and application of six-point linear multistep numerical method for the solution of second order initial value problems which was implemented in block mode. Yusuph and Onumanyi (2002) demonstrated a successful application of multiple finite difference methods through multistep collocation for the second order ordinary differential equations.

Furthermore, Abualnaja (2015) constructed a block procedure with linear multistep methods using Legendre polynomial for solving first order ordinary differential equations. The method depends on the perturbed collocation approximation with Legendre as perturbation term for the solution of first order ordinary differential equations. In the work of Yakusak *et al.*, (2015), uniform order Legendre approach for continuous hybrid block methods were proposed for the solution of first order ordinary differential equations.

In this paper, we propose the choice of Legendre polynomial without perturbation as basis functions for the construction of continuous schemes, which simultaneously generate solution of (1). They are self-starting and do not need any predictors.

Preliminaries

A central notion in this work concerns the choice of Legendre polynomial as basis functions in the derivation of the continuous schemes, the implementation strategies employed (in block mode) and the stability analysis of the methods. For convenience of the reader, we recall the definitions as follows:

Definition 1.1: The block method is said to be zero-stable if the roots λ_j, $j = 1, 2, \ldots s$ of the characteristic polynomial $\rho(\lambda)$ defined by $\rho(\lambda) = |\sum_{i=0}^{s} A^i \lambda^{s-i}| = 0$ satisfies $|\lambda_j| \leq 1$ and for those roots with $|\lambda_j| = 1$, the multiplicity must not exceed the order of the differential equation. (see Fatunla (1994)).

Definition 1.2: The set of W equals

$\tau \in C$; all roots $\xi_i(\tau)$ of the characteristic equation

satisfy $|\xi_i(\tau)|$, multiple roots satisfy $|\xi_i(\tau)| < 1$

is called the stability region or region of absolute stability of the method (Hairer and Wanner (1996)).

Definition: 1.3 (Widlund (1967))

A method is said to be A(α)-stable if the sector

$S_\alpha = \{z : |\arg(-z)| \leq \alpha, z \neq 0\}$ is contained in the stability region.

Definition 1.4 (Ehle (1969)): A method is called L-stable if it is A-stable and if in addition $\lim_{z \to \infty} R(z) = 0$.

Definition 1.5 (Olagunju *et al.*, (2012)): Legendre polynomial is special case of the Legendre function which satisfy the differential equation

$$(1 - x^2)y'' - 2xy' + n(n+1)y = 0, \quad n > o, \quad |x| < 1.$$

The general solution can be expressed as:

$$y = AP_n(x) + BQ_n(x), \quad |x| < 1.$$

$P_n(x)$ and $Q_n(x)$ are respectively the Legendre functions of the first-and second-kind of the order n. The nth order polynomial $P_n(x)$ is generally given by the following equation:

$$P_n(x) = \sum_{k=0}^{\lfloor \frac{n}{2} \rfloor} (-1)^k \left(\frac{n!}{2n!} \right) \frac{(2n-2k)!}{k!(n-k)!(n-2k)!} x^{n-2k}$$

where n is the order of the Legendre polynomials, $\lfloor \frac{n}{2} \rfloor$ signifies the integer part of $\frac{n}{2}$. Legendre polynomials are orthogonal to each other with respect to weight function w(x) = 1 on [-1,1]. The first two polynomials are always the same in all cases but the higher orders are created with recursive formula:

$$(n+1)P_{n+1}(x) = (2n+1)xP_n(x) - nP_{n-1}(x), \quad n = 1, 2, ...$$

with initial conditions: $P_0(x) = 1, P_1(x) = x$. The first four terms of the polynomial are; $P_0(x) = 1, \quad P_1(x) = x, \quad P_2(x) = \frac{1}{2}(3x^2 - 1), \quad P_3(x) = \frac{1}{2}(5x^3 - 3x)$

The paper is organized as follows. Section 1 is of an introductory nature. The materials and methods are described in Section 2. Stability analysis of the methods is discussed in Section 3. In Section 4, some numerical experiments and results showing the relevance of the new methods are discussed. Finally, in Section 5 some conclusions are drawn.

MATHEMATICAL FORMULATION

Consider the second-order initial value problem:

$$y'' = f(x, y, y'), y(a) = \eta_0, y'(a) = \eta_1 \tag{1}$$

where $f \in R$ is sufficiently differentiable and satisfies a Lipschitz condition, sufficiently smooth, $f: \mathbb{R}^{m+1} \to \mathbb{R}^m$, y is an m-dimensional vector and x is a scalar variable and a set of equally spaced points on the integration interval also given by

$$a = x_0 < x_1 < ... < x_n < \cdots < x_{n+k} < x_N = b, \tag{2}$$

with a specified positive integer step number k greater than zero, h can be variable or constant step-size given by

$$h = x_{n+1} - x_n, n = 1, ..., N; hN = b - a.$$

Assuming an approximate solution to (1) by taking the partial sum of Legendre polynomial of the form:

$$y(x) = \sum_{r=0}^{t+s-1} a_r P_r(x), x_n \leq x \leq x_{n+r} \tag{3}$$

where can be used only after certain transformation. The second derivative of (3) gives

$$y''(x) = \sum_{r=0}^{t+s-1} a_r P_r''(x) \tag{4}$$

Substituting (4) into (1) gives

$$\sum_{r=0}^{t+s-1} a_r P_r''(x) = f(x, y(x), y'(x)), x_n \le x \le x_{n+r} \tag{5}$$

where $P_r(x)$ is the Legendre polynomial of degree r, valid in $x_n \le x \le x_{n+r}$ and a_r's are real unknown parameters to be determined and $(t+s-1)$ is the sum number of collocation and interpolation points. The well-known Legendre polynomials are defined on the interval [-1,1].

Derivation of the Continuous Hybrid Multistep Methods

Our objective here is to construct a continuous formulation of the general linear multistep method $\overline{y}(x)$ of degree $r = t+s-1$ $t > 0, s > 0$. Two cases were considered one-step and two-step methods.

CASE 1: One-Step Continuous Hybrid Multistep Method (OSCHMM).

Collocating (5) at points, and interpolating (3) at points $x = x_{n-s}, s = 0, \frac{1}{4}$, respectively lead to a system of equations expressed in matrix form:

$$MD = U \tag{6}$$

where,

$$M = \begin{pmatrix} 1 & -1 & 1 & -1 & 1 & -1 & 1 \\ 1 & -\dfrac{1}{2} & -\dfrac{1}{8} & \dfrac{7}{16} & -\dfrac{37}{128} & -\dfrac{23}{256} & \dfrac{331}{1024} \\ 0 & 0 & 12 & -60 & 180 & -420 & 840 \\ 0 & 0 & 12 & -30 & \dfrac{45}{2} & \dfrac{105}{4} & -\dfrac{2415}{32} \\ 0 & 0 & 12 & 0 & -30 & 0 & \dfrac{105}{2} \\ 0 & 0 & 12 & 30 & \dfrac{45}{2} & -\dfrac{105}{4} & -\dfrac{2415}{32} \\ 0 & 0 & 12 & 60 & 180 & 420 & 840 \end{pmatrix}$$

$D = [a_0, a_1, a_2, a_3, a_4, a_5, a_6,]^T$ and $U = [y_n, y_{n+1/4}, f_n, f_{n+1/4}, f_{n+1/2}, f_{n+3/4}, f_{n+1}]^T$.

Solving (6) using Gaussian Elimination method in Maple soft environment produces the following values of $a_r's$:

$$a_0 = -y_n + 2y_{n+\frac{1}{4}} + h^2 \left(\frac{407}{80640} f_n + \frac{131}{4480} f_{n+\frac{1}{4}} + \frac{1231}{20160} f_{n+\frac{1}{2}} + \frac{181}{20160} f_{n+\frac{3}{4}} - \frac{13}{80640} f_{n+1} \right)$$

(7)

$$a_1 = -2y_n + 2y_{n+\frac{1}{4}} + h^2 \left(\frac{727}{80640} f_n + \frac{37}{640} f_{n+\frac{1}{4}} + \frac{1999}{20160} f_{n+\frac{1}{2}} + \frac{437}{20160} f_{n+\frac{3}{4}} - \frac{13}{80640} f_{n+1} \right)$$

(8)

$$a_2 = h^2 \left(\frac{1}{1512} f_n + \frac{4}{189} f_{n+\frac{1}{4}} + \frac{5}{126} f_{n+\frac{1}{2}} + \frac{4}{126} f_{n+\frac{3}{4}} - \frac{1}{1512} f_{n+1} \right)$$

(9)

$$a_3 = h^2 \left(-\frac{1}{1080} f_n + \frac{4}{189} f_{n+\frac{1}{4}} - \frac{2}{135} f_{n+\frac{3}{4}} + \frac{1}{1080} f_{n+1} \right)$$

(10)

$$a_4 = h^2 \left(\frac{13}{13860} f_n + \frac{4}{693} f_{n+\frac{1}{4}} - \frac{31}{2310} f_{n+\frac{1}{2}} + \frac{4}{693} f_{n+\frac{3}{4}} - \frac{13}{13860} f_{n+1} \right)$$

(11)

$$a_5 = h^2 \left(-\frac{1}{945} f_n + \frac{2}{945} f_{n+\frac{1}{4}} - \frac{2}{945} f_{n+\frac{3}{4}} - \frac{1}{945} f_{n+1} \right)$$

(12)

$$a_6 = h^2 \left(\frac{4}{10395} f_n - \frac{16}{10395} f_{n+\frac{1}{4}} + \frac{8}{3465} f_{n+\frac{1}{2}} + -\frac{16}{10395} f_{n+\frac{3}{4}} + \frac{16}{10395} f_{n+1} \right)$$

(13)

Substituting (7)-(13) into equation (3) and after some manipulation gives the continuous scheme

$$\bar{y}(x) = \left(1 - 4\left(\frac{x-x_n}{h}\right)\right) y_n + 4\left(\frac{x-x_n}{h}\right) y_{n+\frac{1}{4}}$$

$$+ \frac{h^2}{5760} \left(-367\left(\frac{x-x_n}{h}\right) + 2880\left(\frac{x-x_n}{h}\right)^2 - 8000\left(\frac{x-x_n}{h}\right)^3 + 11200\left(\frac{x-x_n}{h}\right)^4 - 7680\left(\frac{x-x_n}{h}\right)^5 + 2048\left(\frac{x-x_n}{h}\right)^6 \right) f_n$$

$$- \frac{h^2}{1440} \left(135\left(\frac{x-x_n}{h}\right) - 3840\left(\frac{x-x_n}{h}\right)^3 + 8320\left(\frac{x-x_n}{h}\right)^4 - 6912\left(\frac{x-x_n}{h}\right)^5 + 2048\left(\frac{x-x_n}{h}\right)^6 \right) f_{n+\frac{1}{4}}$$

(14)

Evaluating (14) at $x = x_{n+1}$ gives the discrete scheme

$$y_{n+1} = 4y_{n+\frac{1}{4}} - 3y_n + \frac{h^2}{1920} \left(27 f_n + 332 f_{n+\frac{1}{4}} + 222 f_{n+\frac{1}{2}} + 132 f_{n+\frac{3}{4}} + 7 f_{n+1} \right)$$

(15)

The discrete scheme (15) is consistent, zero-stable and of order $p = 5$ with error constant $C_{p-2} = \frac{1}{1966080}$.

Here, it is our intention to get additional discrete schemes, so, we evaluated (14) at the points $x = x_{n+i}, i = \frac{3}{4}, \frac{1}{2}$ to obtain:

$$y_{n+\frac{3}{4}} = 3y_{n+\frac{1}{4}} - 2y_n + \frac{h^2}{3840}\left(37f_n + 432f_{n+\frac{1}{4}} + 222f_{n+\frac{1}{2}} + 32f_{n+\frac{3}{4}} - 3f_{n+1} \right) \tag{16}$$

$$y_{n+\frac{1}{2}} = 4y_{n+\frac{1}{4}} - 4y_n + \frac{h^2}{5760}\left(367f_n + 540f_{n+\frac{1}{4}} - 282f_{n+\frac{1}{2}} + 116f_{n+\frac{3}{4}} - 21f_{n+1} \right) \tag{17}$$

The first derivative of (14) is found and evaluated at points $x = x_{n+i}, i = 0, \frac{1}{4}, \frac{1}{2}, \frac{3}{4}, 1$ yields the following derivative schemes:

$$hy'_n = 4y_{n+\frac{1}{4}} - 4y_n - \frac{h^2}{5760}\left(367f_n + 540f_{n+\frac{1}{4}} - 282f_{n+\frac{1}{2}} + 116f_{n+\frac{3}{4}} - 21f_{n+1} \right) \tag{18}$$

$$hy'_{n+\frac{1}{4}} = 4y_{n+\frac{1}{4}} - 4y_n + \frac{h^2}{5760}\left(135f_n + 752f_{n+\frac{1}{4}} - 246f_{n+\frac{1}{2}} + 96f_{n+\frac{3}{4}} - 17f_{n+1} \right) \tag{19}$$

$$hy'_{n+\frac{1}{2}} = 4y_{n+\frac{1}{4}} - 4y_n + \frac{h^2}{5760}\left(97f_n + 1444f_{n+\frac{1}{4}} + 666f_{n+\frac{1}{2}} - 52f_{n+\frac{3}{4}} + 5f_{n+1} \right) \tag{20}$$

$$hy'_{n+\frac{1}{4}} = 4y_{n+\frac{1}{4}} - 4y_n + \frac{h^2}{5760}\left(81f_n + 1508f_{n+\frac{1}{4}} + 1050f_{n+\frac{1}{2}} + 1932f_{n+\frac{3}{4}} + 469f_{n+1} \right) \tag{21}$$

$$hy'_{n+1} = 4y_{n+\frac{1}{4}} - 4y_n + \frac{h^2}{5760}\left(81f_n + 1508f_{n+\frac{1}{4}} + 1050f_{n+\frac{1}{2}} + 1932f_{n+\frac{3}{4}} + 469f_{n+1} \right) \tag{22}$$

Implementation of the One-Step Continuous Hybrid Multistep Method (OSCHMM).

In this section, the implementation strategy of this work is discussed. Following Fatunla (1991, 1994), the general discrete block formula is given as:

$$A^0 Y_m = ey_n + h^\mu dF(Y_m) + h^\mu BF(Y_m) \tag{23}$$

where e, d are vectors, B are RxR matrix and A^0 identity matrix, μ is the order of differential equation. Expressing equations (20) - (23) and (24) in form of (23) and solving with matrix inversion method gives:

$$
\begin{pmatrix} 1 & 0 & 0 & 0 \\ 0 & 1 & 0 & 0 \\ 0 & 0 & 1 & 0 \\ 0 & 0 & 0 & 1 \end{pmatrix}
\begin{pmatrix} y_{n+\frac{1}{4}} \\ y_{n+\frac{1}{2}} \\ y_{n+\frac{3}{4}} \\ y_{n+1} \end{pmatrix}
=
\begin{pmatrix} 0 & 0 & 0 & 1 \\ 0 & 0 & 0 & 1 \\ 0 & 0 & 0 & 1 \\ 0 & 0 & 0 & 1 \end{pmatrix}
\begin{pmatrix} y_{n-\frac{1}{4}} \\ y_{n-\frac{1}{2}} \\ y_{n-\frac{3}{4}} \\ y_n \end{pmatrix}
+ h
\begin{pmatrix} 0 & 0 & 0 & \frac{1}{4} \\ 0 & 0 & 0 & \frac{1}{2} \\ 0 & 0 & 0 & \frac{3}{4} \\ 0 & 0 & 0 & 1 \end{pmatrix}
\begin{pmatrix} y'_{n-\frac{1}{4}} \\ y'_{n-\frac{1}{2}} \\ y'_{n-\frac{3}{4}} \\ y'_n \end{pmatrix}
+
$$

$$
\tag{24}
$$

$$
h^2
\begin{pmatrix} 0 & 0 & 0 & \frac{367}{23040} \\ 0 & 0 & 0 & \frac{53}{1440} \\ 0 & 0 & 0 & \frac{147}{2560} \\ 0 & 0 & 0 & \frac{7}{90} \end{pmatrix}
\begin{pmatrix} f_{n-\frac{1}{4}} \\ f_{n-\frac{1}{2}} \\ f_{n-\frac{3}{4}} \\ f_n \end{pmatrix}
+ h^2
\begin{pmatrix} \frac{3}{128} & -\frac{47}{3840} & \frac{29}{5760} & -\frac{7}{7680} \\ \frac{1}{10} & -\frac{1}{48} & \frac{1}{90} & -\frac{1}{480} \\ \frac{117}{640} & \frac{27}{1280} & \frac{3}{128} & -\frac{9}{2560} \\ \frac{4}{15} & \frac{1}{15} & \frac{4}{45} & 0 \end{pmatrix}
\begin{pmatrix} f_{n+\frac{1}{4}} \\ f_{n+\frac{1}{2}} \\ f_{n+\frac{3}{4}} \\ f_{n+1} \end{pmatrix}
$$

Writing (24) explicitly

$$
y_{n+\frac{1}{4}} = y_n - \frac{h}{4}y'_n + \frac{h^2}{23040}\left(367f_n + 540f_{n+\frac{1}{4}} - 282f_{n+\frac{1}{2}} + 116f_{n+\frac{3}{4}} - 21f_{n+1} \right)
\tag{25}
$$

$$
y_{n+\frac{1}{2}} = y_n - \frac{h}{2}y'_n + \frac{h^2}{1440}\left(53f_n + 144f_{n+\frac{1}{4}} - 30f_{n+\frac{1}{2}} + 16f_{n+\frac{3}{4}} - 3f_{n+1} \right)
\tag{26}
$$

$$
y_{n+\frac{3}{4}} = y_n - \frac{3h}{4}y'_n + \frac{3h^2}{2560}\left(49f_n + 156f_{n+\frac{1}{4}} + 18f_{n+\frac{1}{2}} + 20f_{n+\frac{3}{4}} - 3f_{n+1} \right)
\tag{27}
$$

$$
y_{n+1} = y_n - hy'_n + \frac{h^2}{90}\left(7f_n + 24f_{n+\frac{1}{4}} + 6f_{n+\frac{1}{2}} + 8f_{n+\frac{3}{4}} \right)
\tag{28}
$$

The block method is of uniform order $p = (5, \ 5, \ 5, \ 5)^T$ with error constant
$C_{p+2} = \left(\frac{107}{165150720}, \ \frac{1}{645120}, \ \frac{9}{3670016}, \ \frac{1}{322560} \right)^T.$

Substituting (25) into (19)-(22) yields

$$y'_{n+\frac{1}{4}} = y'_n + \frac{h}{2880}\left(251f_n + 646f_{n+\frac{1}{4}} - 264f_{n+\frac{1}{2}} + 106f_{n+\frac{3}{4}} - 19f_{n+1}\right)$$
(29)

$$y'_{n+\frac{1}{2}} = y'_n + \frac{h}{360}\left(29f_n + 124f_{n+\frac{1}{4}} + 24f_{n+\frac{1}{2}} + 4f_{n+\frac{3}{4}} - f_{n+1}\right)$$
(30)

$$y'_{n+\frac{3}{4}} = y'_n + \frac{3h}{320}\left(9f_n + 34f_{n+\frac{1}{4}} + 24f_{n+\frac{1}{2}} + 14f_{n+\frac{3}{4}} - f_{n+1}\right)$$
(31)

$$y'_{n+1} = y'_n + \frac{h}{90}\left(7f_n + 32f_{n+\frac{1}{4}} + 12f_{n+\frac{1}{2}} + 32f_{n+\frac{3}{4}} + 7f_{n+1}\right)$$
(32)

Equations (25)-(32) are then applied in block form as simultaneous numerical integrators to solve (1).

CASE 2: Two-Step Continuous Hybrid Multistep Method (TSCHMM).

Similarly, collocating (5) at points, and interpolating (3) at points $x = x_{n+s}, s = 0, \frac{1}{2}$ lead to a system of equations of form (6) where,

$$M = \begin{pmatrix} 1 & -1 & 1 & -1 & 1 & -1 & 1 \\ 1 & -\dfrac{1}{2} & \dfrac{1}{8} & -\dfrac{7}{16} & \dfrac{37}{128} & -\dfrac{23}{256} & \dfrac{331}{1024} \\ 0 & 0 & 3 & -15 & 45 & -105 & 210 \\ 0 & 0 & 3 & -\dfrac{15}{2} & \dfrac{45}{8} & \dfrac{105}{16} & -\dfrac{2415}{128} \\ 0 & 0 & 3 & 0 & -\dfrac{15}{2} & 0 & \dfrac{105}{8} \\ 0 & 0 & 3 & \dfrac{15}{2} & \dfrac{45}{8} & -\dfrac{105}{16} & -\dfrac{2415}{128} \\ 0 & 0 & 3 & 15 & 45 & 105 & 210 \end{pmatrix}$$

$D = [a_0, a_1, a_2, a_3, a_4, a_5, a_6,]^T$ and $U = [y_n, y_{n+1/2}, f_n, f_{n+1/2}, f_{n+1}, f_{n+3/2}, f_{n+2}]^T$. The continuous scheme is as follows:

$$\overline{y}(x) = \left(1 - 2\left(\frac{x-x_n}{h}\right)\right)y_n + 2\left(\frac{x-x_n}{h}\right)y_{n+\frac{1}{4}}$$

$$+ \frac{h^2}{210240}\left[-26791\left(\frac{x-x_n}{h}\right) + 105120\left(\frac{x-x_n}{h}\right)^2 - 14400\left(\frac{x-x_n}{h}\right)^3 + 102200\left(\frac{x-x_n}{h}\right)^4 - 35040\left(\frac{x-x_n}{h}\right)^5 + 4672\left(\frac{x-x_n}{h}\right)^6\right]f_n$$

$$- \frac{h^2}{720}\left[135\left(\frac{x-x_n}{h}\right) - 960\left(\frac{x-x_n}{h}\right)^3 + 1040\left(\frac{x-x_n}{h}\right)^4 - 432\left(\frac{x-x_n}{h}\right)^5 + 64\left(\frac{x-x_n}{h}\right)^6\right]f_{n+\frac{1}{2}}$$

(33)

$$+ \frac{h^2}{480}\left[47\left(\frac{x-x_n}{h}\right) - 480\left(\frac{x-x_n}{h}\right)^3 + 760\left(\frac{x-x_n}{h}\right)^4 - 384\left(\frac{x-x_n}{h}\right)^5 + 64\left(\frac{x-x_n}{h}\right)^6\right]f_{n+1}$$

$$+ \frac{h^2}{720}\left[29\left(\frac{x-x_n}{h}\right) - 320\left(\frac{x-x_n}{h}\right)^3 + 560\left(\frac{x-x_n}{h}\right)^4 - 336\left(\frac{x-x_n}{h}\right)^5 + 64\left(\frac{x-x_n}{h}\right)^6\right]f_{n+\frac{3}{2}}$$

$$+ \frac{h^2}{2880}\left[21\left(\frac{x-x_n}{h}\right) - 2880\left(\frac{x-x_n}{h}\right)^3 + 440\left(\frac{x-x_n}{h}\right)^4 - 288\left(\frac{x-x_n}{h}\right)^5 + 64\left(\frac{x-x_n}{h}\right)^5\right]f_{n+2}$$

The discrete scheme is obtained as:

$$y_{n+2} = 4y_{n+\frac{1}{2}} - 3y_n + \frac{h^2}{1920}\left(27f_n + 332f_{n+\frac{1}{2}} + 222f_{n+1} + 132f_{n+\frac{3}{2}} + 7f_{n+2}\right)$$

The discrete scheme (34) is consistent, zero-stable and of order $p = 5$, with the error constant $C_{p-2} = \frac{1}{15360}$.

Evaluating (33) at the points $x = x_{n-i}, i = \frac{3}{2}, 1$, we obtained the following discrete schemes:

$$y_{n+\frac{3}{2}} = 3y_{n+\frac{1}{2}} - 2y_n + \frac{h^2}{3840}\left(37f_n + 432f_{n+\frac{1}{2}} + 222f_{n+1} + 32f_{n+\frac{3}{2}} - 3f_{n+2}\right)$$

(35)

$$y_{n+1} = 2y_{n+\frac{1}{2}} - y_n + \frac{h^2}{5760}\left(19f_n + 204f_{n+\frac{1}{2}} + 14f_{n+1} + 4f_{n+\frac{3}{2}} - f_{n+2}\right)$$

(36)

The first derivative of (33) is found and evaluated at points $x = x_{n-i}, i = 0, \frac{1}{2}, 1, \frac{3}{2}, 2$ yield the following derivative schemes:

$$hy'_n = 2y_{n+\frac{1}{2}} - 2y_n - \frac{h^2}{2880}\left(367f_n + 540f_{n+\frac{1}{2}} - 282f_{n+1} + 116f_{n+\frac{3}{2}} - 21f_{n+2}\right)$$

(37)

$$hy'_{n+\frac{1}{2}} = 2y_{n+\frac{1}{2}} - 2y_n + \frac{h^2}{2880}\left(135f_n + 752f_{n+\frac{1}{2}} - 246f_{n+1} + 96f_{n+\frac{3}{2}} - 17f_{n+2}\right)$$
(38)

$$hy'_{n+1} = 4y_{n+\frac{1}{2}} - 4y_n + \frac{h^2}{2880}\left(97f_n + 1444f_{n+\frac{1}{2}} + 666f_{n+1} - 52f_{n+\frac{3}{2}} + 5f_{n+2}\right)$$
(39)

$$hy'_{n+\frac{3}{2}} = 2y_{n+\frac{1}{2}} - 2y_n + \frac{h^2}{2880}\left(119f_n + 1296f_{n+\frac{1}{2}} + 1578f_{n+1} + 640f_{n+\frac{3}{2}} - 33f_{n+2}\right)$$
(40)

$$hy'_{n+1} = 2y_{n+\frac{1}{2}} - 2y_n + \frac{h^2}{2880}\left(81f_n + 1508f_{n+\frac{1}{2}} + 1050f_{n+1} + 1932f_{n+\frac{3}{2}} + 469f_{n+2}\right)$$
(41)

The implementation of Two-Step Continuous Hybrid Multistep Method (TSCHMM) is as follows:

Combining (34), (35), (36), (37) and solving by using matrix inversion method gives:

$$
\begin{pmatrix} 1 & 0 & 0 & 0 \\ 0 & 1 & 0 & 0 \\ 0 & 0 & 1 & 0 \\ 0 & 0 & 0 & 1 \end{pmatrix}
\begin{pmatrix} y_{n+\frac{1}{2}} \\ y_{n+1} \\ y_{n+\frac{3}{2}} \\ y_{n+2} \end{pmatrix}
=
\begin{pmatrix} 0 & 0 & 0 & 1 \\ 0 & 0 & 0 & 1 \\ 0 & 0 & 0 & 1 \\ 0 & 0 & 0 & 1 \end{pmatrix}
\begin{pmatrix} y_{n-\frac{1}{2}} \\ y_{n-1} \\ y_{n-\frac{3}{2}} \\ y_n \end{pmatrix}
+ h
\begin{pmatrix} 0 & 0 & 0 & \frac{1}{2} \\ 0 & 0 & 0 & 1 \\ 0 & 0 & 0 & \frac{3}{2} \\ 0 & 0 & 0 & 2 \end{pmatrix}
\begin{pmatrix} y'_{n-\frac{1}{2}} \\ y'_{n-1} \\ y'_{n-\frac{3}{2}} \\ y'_n \end{pmatrix}
+
$$

$$
h^2
\begin{pmatrix} 0 & 0 & 0 & \frac{367}{57600} \\ 0 & 0 & 0 & \frac{53}{360} \\ 0 & 0 & 0 & \frac{147}{640} \\ 0 & 0 & 0 & \frac{14}{45} \end{pmatrix}
\begin{pmatrix} f_{n-\frac{1}{2}} \\ f_{n-1} \\ f_{n-\frac{3}{2}} \\ f_n \end{pmatrix}
+ h^2
\begin{pmatrix} \frac{3}{32} & -\frac{47}{960} & \frac{29}{1440} & -\frac{7}{1920} \\ \frac{2}{5} & \frac{1}{12} & \frac{2}{45} & -\frac{1}{120} \\ \frac{117}{160} & \frac{27}{320} & \frac{3}{32} & -\frac{9}{640} \\ \frac{16}{15} & \frac{4}{15} & \frac{16}{45} & 0 \end{pmatrix}
\begin{pmatrix} f_{n+\frac{1}{2}} \\ f_{n+1} \\ f_{n+\frac{3}{2}} \\ f_{n+2} \end{pmatrix}
$$
(42)

The above block method is of uniform order $p = (5, 5, 5, 5)^T$ with the error constant $C_{p+2} = \left(\frac{107}{1290240}, \frac{1}{5040}, \frac{9}{28672}, \frac{1}{2520}\right)^T$. Equations (38)-(41) with (42) are applied in block form as simultaneous numerical integrators to solve (1).

STABILITY ANALYSIS

In the spirit of Sommeijer et al., (1992), the linear stability of block method can be investigated by applying the method to the test equation $y'' = \lambda y$. This leads to a recursion of the form:

$$Y_{n+2} = M(z)Y_n,$$
$$M(z) := [I - zD]^{-1}[A + zB], \quad z := \lambda h$$

M is called the amplification matrix and its eigenvalues the amplification factors. By requiring the elements of the diagonal matrix D to be positive, the matrix $I - zD$ is nonsingular for all z on the negative real axis. Therefore, in the sequel, we assume that the (diagonal) elements of D are positive. We shall use the result on the power of a matrix N (Varga 1962),

$$\|N^n\| = o(n^{q-1}[\rho(N)]^n) \quad \text{as} \quad n \to \infty,$$

where and $\rho(N)$ are the spectra norm and the radius of N and where all diagonal sub-matrices of the Jordan normal form of N which have spectral radius $\rho(N)$ are at most qxq. If the spectra radius $\rho(N) < 1$, then N is called power bounded. The region where the amplification matrix M(z) is power bounded is called the stability region of the block method. If the stability region contains the origin, then the method is called the zero-stable. Below are the graphical representations of stability of OSCHMM and TSCHMM respectively.

NUMERICAL EXPERIMENTS AND RESULTS

In this section, we applied the new methods to some problems: the first is Undamped Duffing's equation of Fang and Wu (2008), two body problem of Fatunla (1990), stiff problem, linear second order initial value problem, Stiefel and Bettis problem and Implicit 3-point 1-block (I3P1B) of Fadzial (2009).

Problem 1: The Undamped Duffing's equation:

$$y'' + y = -y^3 + (\cos t + \varepsilon \sin 10t)^3 - 99\varepsilon \sin 10t$$
$$y(t_0) = 1, \quad y'(t_0) = 10\varepsilon, \quad \varepsilon = 10^{-10}.$$

The exact solution $y(t) = \cos t + \varepsilon \sin 10t$.

It describes a periodic motion of low frequency with a small perturbation of high frequency. The numerical results are shown on tables 1 and 2 below.

Table 1: One-step method for problem 1 Undamped Duffing's Equation

X	y-exact solution	y-approximate	Error
0.0025	0.9999968750041270	0.9999968750041270	0.000000e+000
0.0050	0.9999875000310390	0.9999875000310390	1.110223e-016
0.0075	0.9999718751393280	0.9999718751393290	8.881784e-016
0.0100	0.9999500004266480	0.9999500004266490	7.771561e-016
0.0125	0.9999218760297140	0.9999218760297150	4.440892e-016
0.0150	0.9998875021243030	0.9998875021243040	9.992007e-016
0.0175	0.9998468789252480	0.9998468789252500	1.665335e-015
0.0200	0.9998000066864440	0.9998000066864470	2.775558e-015
0.0225	0.9997468857008410	0.9997468857008460	5.440093e-015
0.0250	0.9996875163004430	0.9996875163004500	7.216450e-015
0.0275	0.9996218988563060	0.9996218988563160	9.436896e-015

Table 2: The y-exact, y-approximate and error of TSCHMM for problem 1

Two-step method, h = 0.01 problem 1 Undamped Duffing's Equation			
X	y-exact solution	y-approximate	Error
0.0050	0.9999875000310390	0.9999875000310390	1.110223e-016
0.0100	0.9999500004266480	0.9999500004266540	5.551115e-015
0.0150	0.9998875021243030	0.9998875021243660	6.328271e-014
0.0200	0.9998000066864440	0.9998000066864920	4.740652e-014
0.0250	0.9996875163004430	0.9996875163005230	8.015810e-014
0.0300	0.9995500337785390	0.9995500337786070	6.827872e-014
0.0350	0.9993875625577780	0.9993875625578920	1.136868e-013
0.0400	0.9992001066999190	0.9992001067000950	1.757483e-013
0.0450	0.9989876708913390	0.9989876708916910	3.522738e-013
0.0500	0.9987502604429080	0.9987502604433770	4.684031e-013

Problem 2: Consider the given two-body problem

$$y_1'' = \frac{-y_1}{r}, \quad y_1(0) = 1, y_1'(0) = 0,$$

$$y_2'' = \frac{-y_2}{r}, \quad y_2(0) = 0, y_2'(0) = 1,$$

$$r = \sqrt{\left(y_1^2 + y_2^2\right)} \quad x \in [0, 15\pi]$$

Theoretical solution: $y_1(x) = \cos x; \quad y_2(x) = \sin x$

Problem 3: Consider the stiff problem

$$y'' + 1001y' + 1000y = 0, \quad y(0)=1, y'(0)=1, h = 0.1$$

Theoretical solution: $y(x) = \exp(-x)$.

Problem 4:

$$y'' - 4y' + 8y = x^3, \quad y(0) = 2, \ y'(0) = 4, \ x \in [0,1],$$

Theoretical solution:

$$y(x) = e^{2x}\left(2\cos(2x) - \frac{3}{64}\sin(2x)\right) + \frac{3x}{32} + \frac{3x^2}{16} + \frac{x^2}{8}$$

Problem 5: Consider the system of equations of Stiefel and Bettis problem:

$$y_1'' + y_1 = 0.001\cos(x), y_1(0) = 1, \quad y_1'(0) = 0$$

$$y_2'' + y_2 = 0.001\sin(x), y_2(0) = 1, \quad y_2'(0) = 0.9995,$$

$$x \in [0, 40\pi],$$

The exact solutions are given as:

$$y_1(x) = \cos(x) + 0.0005x\sin x,$$

$$y_2(x) = \sin(x) - 0.0005x\cos x.$$

On tables 3 and 4, y-exact, y-computed and error of OSCHMM and TSCHMM for problem 2 are shown while the y-exact, y-computed and error of TSCHMM for problem 3 is shown on table 5. On table 6, it is observed that the maximum absolute error of the TSCHMM is 7.14260896951E-08 is (smaller) more accurate than 5.10704 E-06 of Jator & Li (2009) for problem 4.

Table 3: The y-exact, y-approximate and error of OSCHMM for problem 2Table 4: The y-exact, y-approximate and error of TSCHMM for problem 2

X	y_1-exact	y_2-exact	y_1- approximate	y_2- approximate	Error in y_1	Error in y_2
0.1	0.9950041652780	0.0998334166468	0.99500416527	0.99833416647	1.2891E-14	3.0923E-13
0.2	0.9800665778412	0.1986693307950	0.98006657784	0.198669330795	5.1308E-14	6.1279E-13
0.3	0.9553364891256	0.2955202066613	0.95533648912	0.295520206662	1.1448E-13	9.0512E-13
0.4	0.9210609940028	0.3894183423086	0.92106099400	0.389418342309	2.0114E-13	1.1807E-12
0.5	0.8775825618903	0.4794255386042	0.87758256189	0.479425538605	3.0954E-13	1.4346E-12
0.6	0.8253356149096	0.5646424733950	0.82533561490	0.564642473396	4.3747E-13	1.6617E-12
0.7	0.7648421872844	0.6442176872376	0.76484218728	0.644217687239	5.8231E-13	1.8576E-12
0.8	0.6967067093471	0.7173560908995	0.69670670934	0.717356090901	7.4105E-13	2.0184E-12
0.9	0.6216099682706	0.7833269096274	0.62160996826	0.783326909629	9.1035E-13	2.1406E-12
1.0	0.5403023058681	0.8414709848078	0.54030230586	0.841470984810	1.0865E-12	2.2211E-12

Table 4: The y-exact, y-approximate and error of TSCHMM for problem 2

X	y_1-exact	y_2-exact	y_1-approximate	y_2-approximate	Error in y_1	Error in y_2
0.1	0.99500416652780	0.09983341664468	0.99500416527	0.99833416666	1.6897E-12	1.9734E-11
0.2	0.98006657778412	0.19866933079350	0.98006657783	0.19866933084	3.2802E-12	3.9280E-11
0.3	0.95533648891256	0.29552020669613	0.95533648911	0.295520206719	8.1501E-12	5.7761E-11
0.4	0.92106099940028	0.38941834230862	0.92106099399	0.389418342843	1.2859E-11	7.5690E-11
0.5	0.87758256189903	0.47942553860042	0.87758256186	0.479425538695	2.0553E-11	9.1547E-11
0.6	0.82533561490096	0.56464247339500	0.82533561488	0.564642473501	2.7968E-11	1.0652E-10
0.7	0.76484218728441	0.64421768723760	0.76484218724	0.644217687356	3.7894E-11	1.1853E-10
0.8	0.69670670093471	0.71735609089950	0.69670670929	0.717356091028	4.7377E-11	1.2941E-10
0.9	0.62160996882706	0.78332690962740	0.62160996821	0.783326909764	5.8742E-11	1.3657E-10
1.0	0.54030230586681	0.84147098480780	0.54030230579	0.841470984950	6.9468E-11	1.4243E-10

Table 5: The y-exact, y-approximate and error of OSCHMM for problem 3

X	y-exact	y-approximate	Error in OSCHMM (of problem 3)
0.1	0.90483741803595957316	0.90483741803591096220	4.861096E-14
0.2	0.81873075307798185867	0.81873075307788970825	9.215042E-14
0.3	0.74081822068171786607	0.74081822068158785625	1.300098E-13
0.4	0.67032004603563930074	0.67032004603547725100	1.620497E-13
0.5	0.60653065971263342360	0.60653065971244499082	1.884327E-13
0.6	0.54881163609402643263	0.54881163609381692607	2.095065E-13
0.7	0.49658530379140951470	0.49658530379118379194	2.257227E-13
0.8	0.44932896411722159143	0.44932896411698400947	2.375819E-13
0.9	0.40656965974059911188	0.40656965974035351527	2.455966E-13
1.0	0.36787944117144232160	0.36787944117119205438	2.502672E-13

Table 6: Accuracy Comparison of TSCHMM for Problem 4

X	y-exact	y-approx (of problem4)	Error in TSCHMM p = 5	Error in Jator & Li (2009), p = 5
0.1	2.3941125769963956181	2.3941125055703059230	7.14260896951E-08	5.10704 E-06
0.2	2.7481413324264235256	2.7481411575175150467	1.74908908478 E-07	1.49586 E-05
0.3	3.0078669405110678859	3.0078665760231918731	3.64487876012 E-07	2.78532 E-05
0.4	3.1017624057742078185	3.1017617867959027691	6.18978305049 E-07	4.28908 E-05
0.5	2.9395431007452620774	2.9395421018471828849	9.98898079192 E-07	6.70307 E-05
0.6	2.4118365344157147255	2.4118350550184086061	1.47939730611 E-06	1.02637 E-04
0.7	1.3915548304898433104	1.3915527282654101737	2.10222443313E-06	1.44907 E-04
0.8	-0.26232675833435763l	-0.2623295992140394222	2.84087968179 E-06	1.90905 E-04
0.9	-2.697771160773070925	-2.6977748297643383106	3.66899126738 E-06	2.39733 E-04
1.0	-6.058560720845666951	-6.0585652825136523363	4.56166798538 E-06	2.94670 E-04

The accuracy comparison of the new methods and I31PB are shown on table 7. The new One-step method (OSCHMM) and Two-step method (TSCHMM) are substantially more accurate than the numerical solution of initial-value problems (IVPs) using I3P1B, as the maximum absolute error is smaller with variable h is 9.115552(-10) while that the maximum absolute error of TSCHMM is 7.294766(-9) which is smaller than 2.14918(-8) of I3P1B for h = 0.01.

Table 7: Accuracy Comparison of OSCHMM and TSCHMM with Implicit 3-point-1 block (13PIB)

H	METHOD	MAX ERR
0.01	13P1B	2.14918(-8)
	OSCHMM	9.115552(-10)
	TSCHMM	7.294766(-9)
0.005	13P1B	1.34949(-9)
	OSCHMM	1.39353(-10)
	TSCHMM	9.11552(-10)
0.001	13P1B	8.64701(-11)
	OSCHMM	9.114593(-13)
	TSCHMM	7.291698(-12)
0.0005	13P1B	3.95277(-10)
	OSCHMM	1.139323(-13)
	TSCHMM	9.114593(-13)

Table 8: y-exact, y-approximate and error in TSCHMM for problem 5

X	y_1-exact	y_2-exact	y_1-approx (prob5)	y_2-approx (TSCHMM)	Error y_1	Error y_2
0.1	0.9999951220742781	0.003123432421368851	0.9999951220742781	0.003123432421368851	1.64E-18	7.20E-21
0.2	0.9999804883447010	0.006246834371010263	0.9999804883447010	0.006246834371010263	2.87E-18	2.40E-21
0.3	0.9999560989540329	0.009370175377494076	0.9999560989540329	0.009370175377494076	1.26E-18	4.33E-20
0.4	0.9999219541402128	0.012493424969984680	0.9999219541402128	0.012493424969984680	5.73E-18	6.30E-20
0.5	0.9998780542363521	0.015616552678538286	0.9998780542363521	0.015616552678538286	4.10E-18	1.09E-19
0.6	0.9998243996707314	0.018739528034400182	0.9998243996707314	0.018739528034400182	8.60E-18	1.15E-19
0.7	0.9997609909667961	0.021862320570301988	0.9997609909667961	0.021862320570301988	6.97E-18	1.85E-19
0.8	0.9996878287431515	0.024984899820758884	0.9996878287431515	0.024984899820758884	1.14E-17	1.81E-19
0.9	0.9996049137135566	0.028107235322366826	0.9996049137135566	0.028107235322366826	9.83E-18	2.79E-19
1.0	0.9995122466869173	0.031229296614099748	0.9995122466869173	0.031229296614099748	1.43E-17	2.61E-19

CONCLUSIONS

We have presented continuous hybrid multistep methods with Legendre polynomial as basis function for the direct solution of system of second order ODEs. The derived methods were implemented in block mode which have the advantages of being self-starting, uniformly of the same order of accuracy and do not need predictors, having good accuracy as shown on table 7. It should be noted that accuracy and efficiency rate of a method is dependent on the implementation strategies. If economical computation is required, then the new methods are the better choice. The new methods are therefore recommended for general purposed use. Finally, the region of absolute stability of the block methods of One-step and Two-step methods were presented in figures 1 and 2. Maple and Matlab software package were employed to generate the schemes and results.

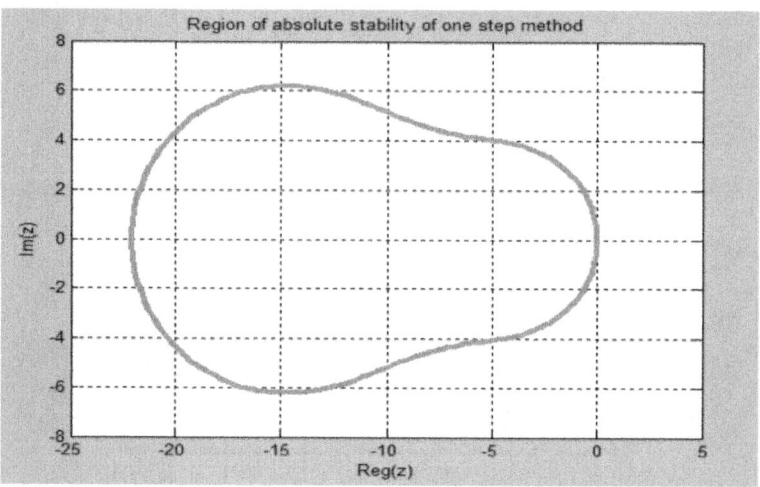

Figure 1: Stability Domain of Block of OSCHMM which is A(α)-**stable by Definition 1.3**

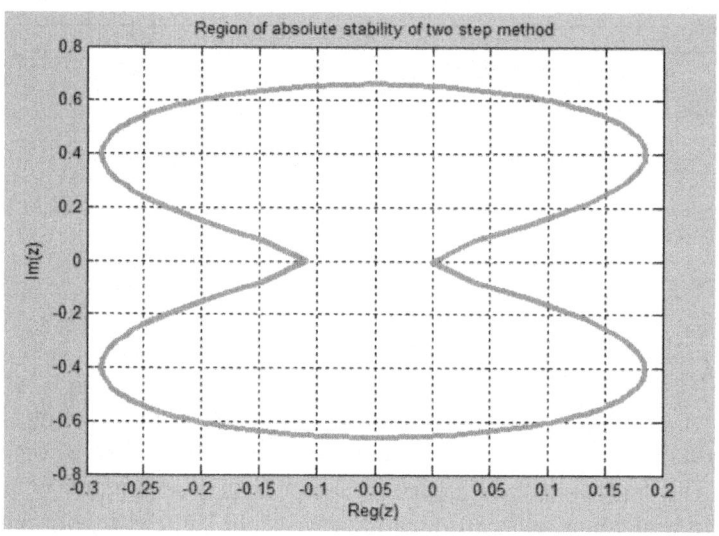

Figure 2: Stability Domain of Block TSCHMM which is L(α)-**stable by definition 1.3 and 1.4**

REFERENCES

1.   Abualnaja, K. M (2015) A Block Procedure with Linear Multistep Methods using Legendre Polynomial for Solving ODEs. Applied Mathematics

SciRes, 6, pp717-723 http://dx.doi.org/10.4236/am.2015.64067.

2.   Adeyefa, E. O. Folaranmi O. R. and A. F Adebisi, (2014), A Self-Starting First Order Initial Value Solver, In. J. Pure Appl. Sc. Technol., 25(1), 8-13.

3.   Awari (2013) Derivation and Application of Six-Point Linear Multistep Numerical Method for the Solution of Second Order Initial Value Problems. IOSR Journal of Mathematics, e-ISSN: 2278-5728, p-ISSN 2319-765X, vol. 7, Issue 2, PP 23-29.

4.   Awoyemi D. O and Idowu, O. M (2005) A Class of Hybrid Collocation Methods for Third Order Ordinary Differential Equations. International Journal of Computer Mathematics vol.82, No 00.pp1-7.

5.   Ehigie, J.O (2010) On Generalized 2-Step Continuous Linear Multistep Method of Hybrid Type For The integration of Second Order Ordinary Differential Equations, Arch. Appl. Sci. Res., 2010,2(6):362-372.

6.   Ehle, B. L. (1969): On Pade Approximations to the Exponential Junction and A-stable Methods for the Numerical Solution of Initial Value Problems. Research Report CSRR 2010, Dept. AACS, Univ. of Waterloo, Ontario, Canada. IV.5.

7.   Fang, Y. L; Songa, Y and Wu, X. Y (2008) A Robust Trigonometrically Fitted Embedded Pair For Perturbed Oscillators. Journal of Computational and Applied Mathematics 225 (2009) 347-355.

8.   Fatunla, S. O (1990) Block Method For Second Order ODEs. Inter. J. Compter Maths, 41(1990), pp55-63.

9.   Fudzial, I ;Yap Lee Ken and Mohamad Othman (2009) Explicit and Implicit 3-Point Block Methods for Solving Special Second Order Ordinary Differential Equations Directly. Int. Journal of Math. Analysis, Vol. 3. No5, 239-254, pp239-253.

10.   Hairer, E and Wanner, G (1996) Solving Ordinary Differential Equations II, Stiff and Differential –Algebraic Problems second revised edition ISBN3-540-60452-9 Springer-Verlag Berlin Heidelberg New York. pp 241.

11.   Jator S. N and Li (2009) A Self-Starting Linear Multistep Method for the Direct Solution of General Second Order Initial Value Problem. International Journal of Computer Math. 86(5), 817-836.

12.   Lambert, J. D. (1991) Numerical Solution for Ordinary Differential Systems, New York John Wiley, 1991.

13.   Landau, L. D and Lifshitz, F. M. (1965) Quantum Mechanics, Pergamon, New York.

14. Liboff, R. L (1980) Introductory Quantum Mechanics, Addison-Wiley, Reading M. A 1980.

15. Olagunju, A.S; Joseph, F. L and Raji, M. T (2012) Comparative Study of The Effect of Different Collocation Point on Ledendre Collocation Methods of Solving Second Order Boundry Value Problems, IOSR Journal of Mathematics vol.7(2) pp 35-41.

16. Onumanyi, P; Oladele, J. O; Adeniyi, R. B and Awoyemi, D. O. (1993) Derivation of Finite Difference Methods by Collocation, Abacus, 23(2), 72-83.

17. Sirisena, U. W; Kumleng, G. M and Yahaya, Y.A (2004) A New Butcher Type Two-Step Block Hybrid Multistep Method For Accurate and Efficient Parallel Solution of ODEs. ABACUS Journal of Mathematical Association of Nigeria. Volume 31, Number 2A Mathematics series NR-ISSN0001-3099 pp1-7.

18. Sommeijer, B. P; Couzy, W. and Van De Houwen (1992) A-Stable Parallel Block Methods for Ordinary and Integro-Differential Equations, Applied Numerical Mathematics 9(1992) 267-281 North-Holland.

19. Yusuph Y and Onumanyi, P (2002) New Multiple FDMS Through Multistep Collocation for $y''= f(x,y)$. Abacus vol. 29 No 2 Mathematics series. The Journal of Mathematical Association of Nigeria pp 92-100.

20. Varga, R. S (1962) Martix Iterative Analysis. Prentice-Hall, Englewood Cliffs, NJ, pp 65.

21. Widlund, O. B (1967): A note on unconditionally stable linear multistep methods. BIT, vol. 7, pp. 65-70.

Chapter 5

ANALYSIS OF COUETTE FLOW OF A NANOFLUID IN AN INCLINED CHANNEL WITH SORET AND DUFOUR EFFECTS

Yusuf A.[1], Aiyesimi Y. M.[1], Jiya M.[1], Okedayo G. [T,2]

[1]Department of Mathematics and Statistics, Federal University of Technology, Minna, Nigeria

[2]Department of Mathematics, Ondo State University of Science and Technology, Okiti-Pupa, Nigeria

ABSTRACT

The solution to the problem of laminar fluid flow in an inclined parallel walls resulting from the movement of the lower wall while the upper wall remain stationary (Coette flow) in a nanofluid with thermal convection, Soret and Dufour effects with radiation has been obtained using the Modified Adomian Decomposition Method for the first time. The model used for the nanofluid was presented in its rectangular coordinate system and incorporates the effect of Brownian motion, and thermophoresis parameter. A similarity solution is presented which depends on the Prandtl number P_r, Lewis number Le, Brownian motion N_b, thermophoresis number N_t, Soret (Sr) number, Dufour (DU) number, and Grashof numbers Gr_T, Gr_c. It is found that increase in Soret number leads to reduction in temperature and nanofraction profiles while increase in Dufour number enhances the temperature and nanofracion profiles. A comparative analysis of present study was carried out with the numerical on Table 1, and it was observed the two method are in good agreement.

INTRODUCTION

From energy saving perspective, improvement of heat transfer performance in systems is very necessary. Low thermal conductivity of conventional heat transfer fluids such as water and oils is a primary limitation in enhancing the performance and the compactness of systems. Solids typically have a higher thermal conductivity than liquids. For example, copper (Cu) has a thermal conductivity 700 times greater than water and 3000 times greater than engine oil. An innovative and novel technique to enhance heat transfer is to use solid particles in the base fluid (i.e. nanofluids) in the range of sizes 10–50 nm. Abu-Nada et al. [1] investigated natural convection heat transfer enhancement in horizontal concentric annuli field by nanofluid. They found that for low Rayleigh numbers, nanoparticles with higher thermal conductivity produce more enhancements in heat transfer. Ellahi [2] studied magnetohydrodynamic (MHD) flow of non-Newtonian nanofluid in a pipe and observed that the MHD parameter decreases the fluid motion and the velocity profile is larger than that of the temperature profile even in the presence of variable viscosities. Free convection heat transfer in a concentric annulus between a cold square and heated elliptic cylinders in the presence of a magnetic field was investigated by Sheikholeslami et al. [3]. They found that enhancement in heat transfer increases as the Hartmann number increases but it decreases with increase of Rayleigh number. Rashidi et al. [4] considered the analysis of the second law of thermodynamics applied to an electrically conducting incompressible nanofluid fluid flowing over a porous rotating disk. They concluded that magnetic rotating disk drives have important applications in heat transfer enhancement in renewable energy systems. Sheikholeslami et al. [5] used heatline analysis to simulate two phase simulation of nanofluid flow and heat transfer. Their results indicated that the average Nusselt number decreases as the buoyancy ratio number increases until it reaches a minimum value and then starts increasing. Sheikholeslami et al. [6] studied the magnetic field effect on CuO–water nanofluid flow and heat transfer in an enclosure which is heated from below. They found that effect of Hartmann number and heat source length is more pronounced at high Rayleigh number. Effect of nanofluid on heat transfer enhancement has been investigated by different authors [7]. Aiyesimi et al [8] considers Hydromagnetic Boudary-Layer Flow of a Nanofluid Past a Stretching Sheet Embedded in a Darcian Porous Medium with Radiation and it was found out that at lower values of Prandtl number, heat able to diffuse more rapidly out of the system.

In a recent paper Aiyesimi et al [9] extended the model of Khan and Pop [10] to analyse and investigate the convective boundary-layer flow of a nanofluid past a stretching sheet with radiation. It was observed that thermal buoyancy and nanofraction buoyancy enhances the fluid velocity, temperature, and nanofraction. It is appropriate to channelize the work of Aiyesimi et al [10] over an inclined channel with Soret and Dufour effects and use the Adomian Decomposition Method (ADM) to obtain the analytical solution of the model.

This work is a new development in the literature in which an analytical solution of a convective boundary-layer flow of a nanofluid past a stretching channel with Soret and Dufour effects is proposed using the Adomian Decomposition Method.

PROBLEM FORMULATION

A steady, two dimensional boundary layer flow of a nanofluid in an inclined channel at angle Θ is considered. It is assumed that the wall located at $y = 0$ stretches with a velocity $u(x, y) = \dfrac{ax}{h}$ (couette flow), while the other wall at $y = h$ (h = is the width of the channel) remain stationary throughout the flow, where a is constant and x is the coordinate measured along the stretching wall. The temperature T and the nanoparticle fraction C have constants values T_0 and C_0 at $y = 0$ and T_h and C_h at $y = h$ respectively. For this application, we will adopt the natural convection with Soret, Dufour, and Radiation effects for the formulation of Khan and Pop [10] and it is governed by the following equations:Continuity equation:

$$\frac{\partial u}{\partial x} + \frac{\partial v}{\partial y} = 0 \tag{1}$$

Momentum equation:

$$u\frac{\partial u}{\partial x} + v\frac{\partial u}{\partial y} = -\frac{1}{\rho_f}\frac{\partial p}{\partial x} + \upsilon\left(\frac{\partial^2 u}{\partial x^2} + \frac{\partial^2 u}{\partial y^2}\right)$$
$$+ g\beta(T - T_h) + g\beta(C - C_h) + \rho g\sin\Theta \tag{2}$$

Energy equation:-

$$u\frac{\partial T}{\partial x} + v\frac{\partial T}{\partial y} = \alpha\left(\frac{\partial^2 T}{\partial x^2} + \frac{\partial^2 T}{\partial y^2}\right)$$
$$+ \tau\left(D_B\left(\frac{\partial C}{\partial x}\frac{\partial T}{\partial x} + \frac{\partial C}{\partial y}\frac{\partial T}{\partial y}\right) + \frac{D_T}{T_\infty}\left(\left(\frac{\partial T}{\partial x}\right)^2 + \left(\frac{\partial T}{\partial y}\right)^2\right)\right)$$
$$+ \frac{D_M K_T}{C_s C_p}\frac{\partial^2 C}{\partial y^2} - \frac{1}{\rho C_p}\frac{\partial q_r}{\partial y} \tag{3}$$

Nanofraction equation:-

$$u\frac{\partial C}{\partial x} + v\frac{\partial C}{\partial y} = D_B\left(\frac{\partial^2 C}{\partial x^2} + \frac{\partial^2 C}{\partial y^2}\right) +$$
$$\left(\frac{D_M K_T}{T_M}\right)\left(\frac{\partial^2 T}{\partial x^2} + \frac{\partial^2 T}{\partial y^2}\right) \tag{4}$$

Subject to the boundary conditions:

$$y = 0 : u = \frac{ax}{h} \; v = 0 \; T = T_0, \; C = C_0$$

$$y = h : u = 0, \; T = T_h, \; C = C_h$$

$$(5)$$

where u and v are the velocity components along the x and y axes respectively, p is the fluid pressure, ρ_f is the density of the base fluid, a is the thermal diffusivity, υ is the kinematic viscousity, k^* is the thermal conductivity, C_p is the specific heat capacity at constant pressure, a is a positive constant, D_b is the Brownian diffusion coefficient, D_T is the thermopheric diffusion coefficient and $\tau = \frac{(\rho c)_p}{(\rho c)_f}$ is the ratio between the effective heat capacity of the fluid with ρ being the density, c is the volumetric volume expansion coefficient and ρ_p is the density of the particles g is the acceleration due to gravity, β is the volumetric coefficient of thermal expansion, q_r is the radiative heat flux T_M is the mean fluid temperature, k_T is the thermal diffusion ratio, D_M is the diffusion ratio, C_s is concentration susceptibility. Following Roseland approximation we have $q_r = -\frac{4\sigma^*}{3\delta} \frac{\partial T^4}{\partial y}$, where σ^* and δ are the Stefan-Boltzmann constant and the mean absorption coefficient respectively. The temperature differences within the fluid is assumed sufficiently small such that T^4 may be expressed as a linear function of Temperature. Expanding T^4 in Taylor's series about T_k and neglecting higher order terms, we get

$$T^4 \cong 4TT_h^3 - 3T_h^4$$

$$(6)$$

therefore, $\dfrac{\partial q_r}{\partial y} = -\dfrac{16\sigma^*}{3\delta} \dfrac{\partial^2 T^4}{\partial y^2}$, Defining the dimensional stream function $(\psi(x,y))$ in the usual way such that $u = \dfrac{\partial \psi}{\partial y}$ and $v = -\dfrac{\partial \psi}{\partial x}$ and following the work of Sheikholeslami [11]:-

$$\eta = \frac{y}{h}, \; \psi = axf(\eta), \; \theta(\eta) = \frac{T - T_h}{T_0 - T_h},$$

$$\text{and} \; \chi(\eta) = \frac{C - C_h}{C_0 - C_h}$$

$$(7)$$

where $\eta, \; f(\eta), \; \theta(\eta), \; \chi(\eta)$ are the dimensionless fluid distance, velocity profile, temperature profile, and nanoparticle concentration. Neglecting the pressure gradient equations (1) to (6) reduces to the following local similarity solution:-

$$f''' + \text{Re } ff'' - \text{Re } f'^2 + Gr_{Tx}\theta + Gr_{Cx}\chi + K = 0 \tag{8}$$

$$\left(1 + \frac{4Ra}{3}\right)\theta'' + \text{Re Pr } f\theta' + \text{Pr } N_b\chi'\theta' \tag{9}$$

$$+ \text{Pr } N_t\theta'^2 + DU \text{ Pr } \chi'' = 0$$

$$\chi'' + L_e f\chi' + L_e S_r\theta'' = 0 \tag{10}$$

with corresponding boundary conditions:

$$f(0) = 0, \ f'(0) = 1, \ \theta(0) = 1, \ \chi(0) = 0,$$

$$f'(1) = 0 \ \theta(1) = 0, \ \chi(1) = 0. \tag{11}$$

in which:

$$Gr_{Tx} = \frac{h^3 g \beta(x)(T_0 - T_v)}{a\upsilon}$$

$$Gr_{Cx} = \frac{h^3 g \beta(x)(C_0 - C_h)}{a^2\upsilon}, \ \text{Re} = \frac{ha}{\upsilon}, \ K = F\sin\Theta$$

$$F = \frac{h^3 \rho g}{a\upsilon x}, \ Ra = \frac{4\sigma^* T_h^3}{\delta k^*}, \ P_r = \frac{\upsilon}{\alpha} \ L_e = \frac{\upsilon}{D_B},$$

$$N_b = \frac{(\rho c)_p D_B (C_0 - C_h)}{(\rho c)_f \upsilon} \ N_t = \frac{(\rho c)_p D_T (T_0 - T_h)}{(\rho c)_f T_h\upsilon}$$

$$DU = \frac{D_M K_T (C_0 - C_h)}{C_S C_p (T_0 - T_h)} \ S_r = \frac{D_M K_T (T_0 - T_h)}{T_M\upsilon(C_0 - C_h)}$$

are the local Thermal Grashof number, local concentration Grashof number, Renold number, Gravitational parameter, Radiation, Prandtl number, Lewis number, Brownian motion parameter, thermophoresis parameter, Dufour number, Soret number, and Schmidt number, respectively. For the momentum equation to have a similarity solution, the parameters Gr_{Tx} and Gr_{Cx} most be constant and not functions of x. This can be met if volumetric coefficient of thermal expansion β is proportional to x. We therefore assume

$$\beta = \beta_0 x \tag{12}$$

The nonlinear coupled differential equations in (8) to (10) with boundary conditions in (11) was solved using the Modified Adomian Decomposition as described by Ebaid and Al-Armani [12].

RESULTS AND DISCUSSION

The nonlinear coupled differential equations (8) to (10) with boundary conditions (11) are solved using the Modfied Adomian Decomposition Methods. In order to assess the accuracy of the present method, we have compared our solution for $\theta(\eta)$ for different values of η at $Gr_T = 0$ and $Gr_c = 0$ with the Numerical method as shown in Table 1. It was observed that the present method is in good agreement with the Numerical-Method.

Table 1: Comparison of Result for $\theta(\eta)$ with the present work for $P_r = 0.1, Gr_T = 0, DU = 0$ and $Gr_c = 0$

η	Numerical	Present Work
0.0	0.9999	1.0000
0.1	0.8988	0.8976
0.2	0.7977	0.7953
0.3	0.6967	0.6931
0.4	0.5960	0.5913
0.5	0.4956	0.4898
0.6	0.3956	0.3889
0.7	0.2960	0.2891
0.8	0.1969	0.1907
0.9	0.0982	0.0941
1	0.0000	0.0000

Figures 1 to 3 shows the effect of thermal Grashof number (Gr_{Tx}) on the velocity profile, temperature and concentration profile and nanofraction profile. It is observed that the thermal Grashof number enhances the velocity profile and the nanofraction profile while the temperature profile reduces as the thermal Grashof number increases.

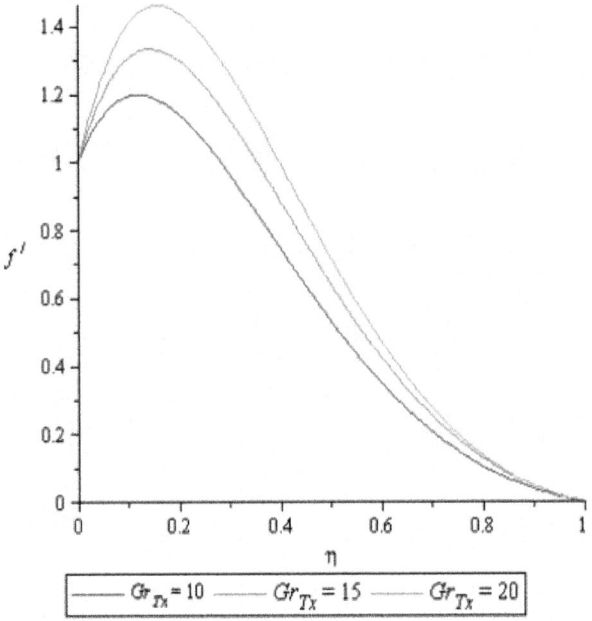

Figure 1: Effect of Gr_{Tx} on Velocity Profile

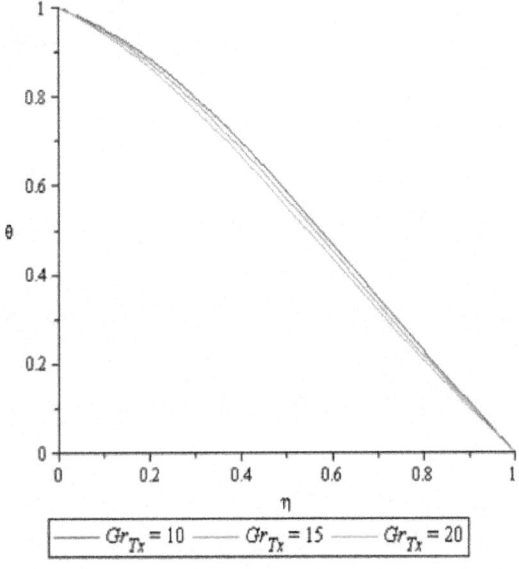

Figure 2: Effect of Gr_{Tx} on Temperature Profile

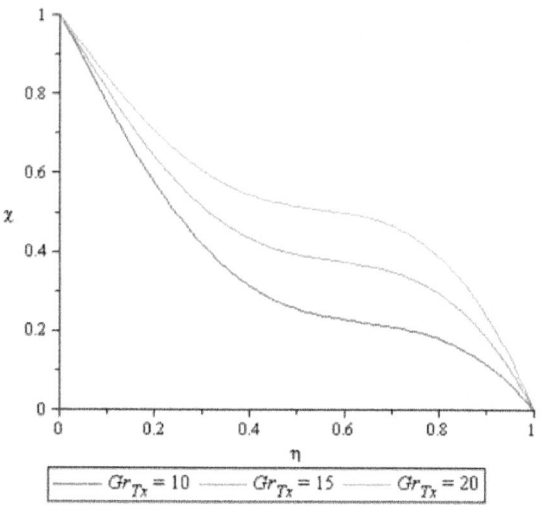

Figure 3: Effect of Gr_{Tx} on Nanofraction Profile

Figure 4 to 6 presents the effect of Concentration Grashof number (Gr_{Cx}). On the velocity profile, temperature and Nanofraction profile. It is observed that the thermal Grashof number enhances the velocity profile at the lower wall (moving wall), where the fluid velocity is maximum and decreases the velocity profile faster for higher values of Concentration profile at the upper wall (stationary wall), where the flow velocity is minimum. The temperature profile and the nanofraction profiles are enhanced. This leads to increase in the boundary layers as shown in the graph.

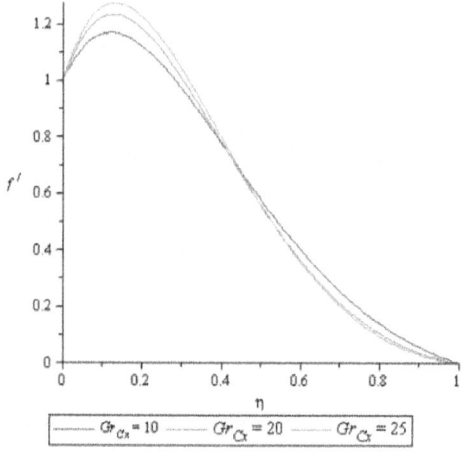

Figure 4: Effect of Gr_{Cx} on Velocity Profile

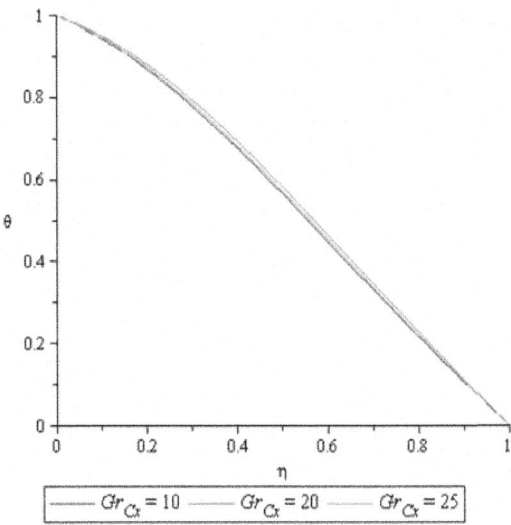

Figure 5: Effect of Gr_{Cx} on Temperature Profile

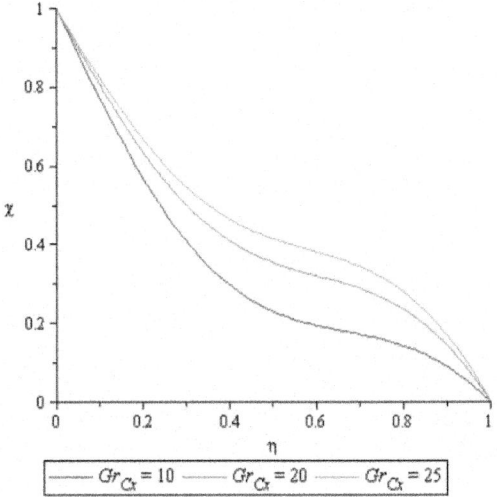

Figure 6: Effect of Gr_{Cx} on Nanofraction Profile

Figures 7 to 8 displays the effect of Prandtl number (Pr) on the temperature profile and the concentration profile. The thermal boundary thickness decreases for both temperature and concentration profile as the Prandtl number decreases. The reason is that higher values of Prandtl number are equivalent to increase in the thermal conductivity of the fluid and therefore heat is able to diffuse away from the heated surface more rapidly for lower values of Prandtl number. Hence there is a reduction in temperature with decrease in the Prandtl number.

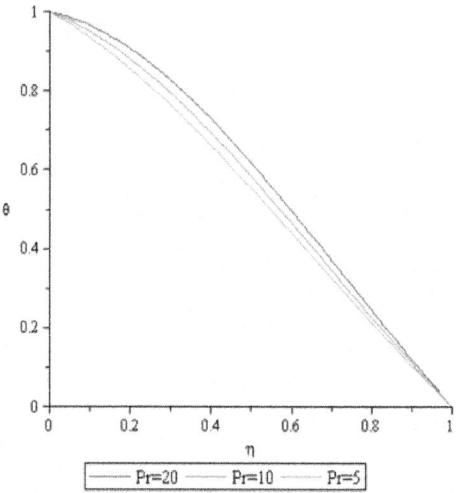

Figure 7: Effect of Pr number on Temperature Profile

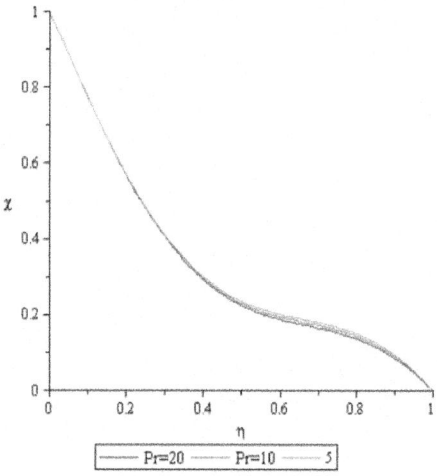

Figure 8: Effect of Pr number on Nanofraction Profile

Figures 9 to 10 depict the effect of Dufour number on temperature and concentration profiles. It is observed that increase in the Dufour number leads to increase in the thermal boundary layer and the concentration boundary layer thickness.

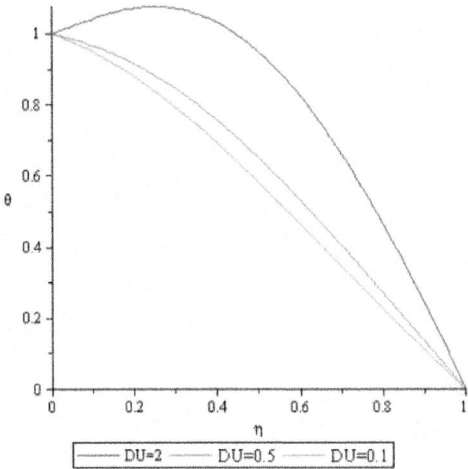

Figure 9: Effect of DU on Temperature Profile

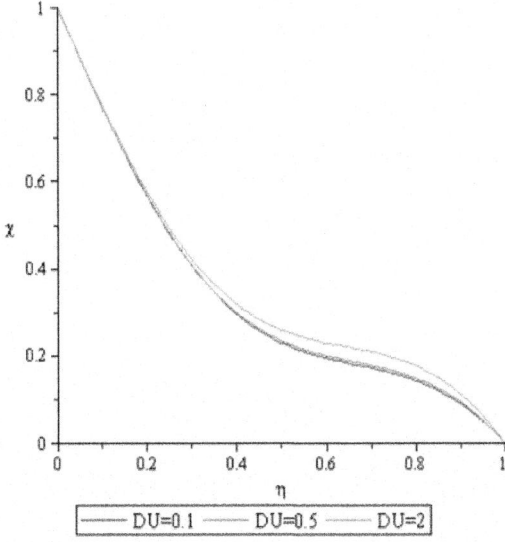

Figure 10: Effect of DU on Nanofraction Profile

Figures 11 to 12 show that the fluid temperature and concentration respectively attains their maximum value at the moving plate surface and decreases monotonically to free stream zero value away from the plate satisfying the boundary conditions. It is observe that increase in radiation (Ra) causes both the temperature and concentration profiles to increase.

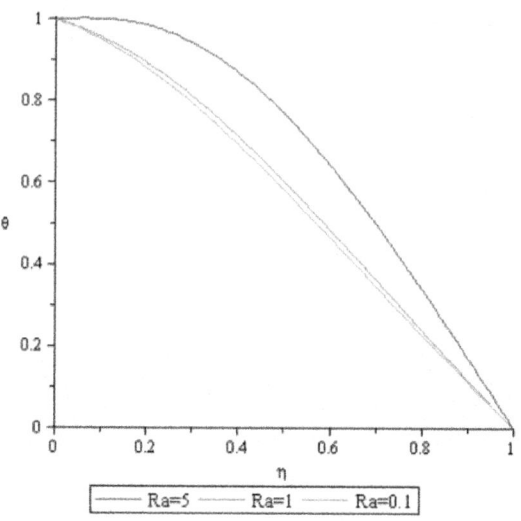

Figure 11: Effect of Ra on Temperature Profile

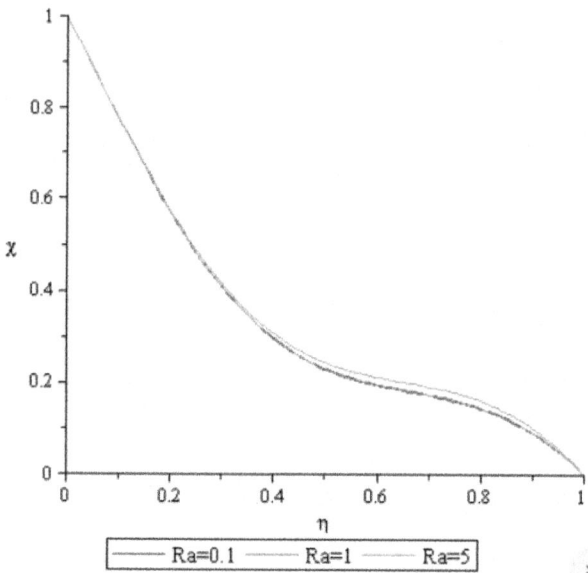

Figure 12: Effect of Ra on Nanofraction Profile

Figures 13 to 14 present the effect of Lewis number (Le) on both the temperature and the concentration profiles respectively. It is observe that increase in Lewis number causes the both the temperature and concentration profiles to reduce.

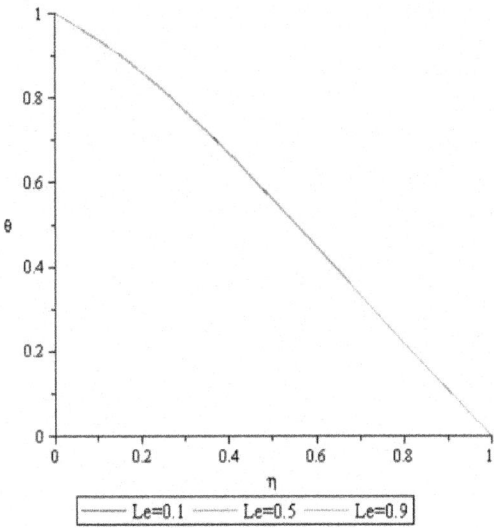

Figure 13: Effect of Le on Temperature Profile

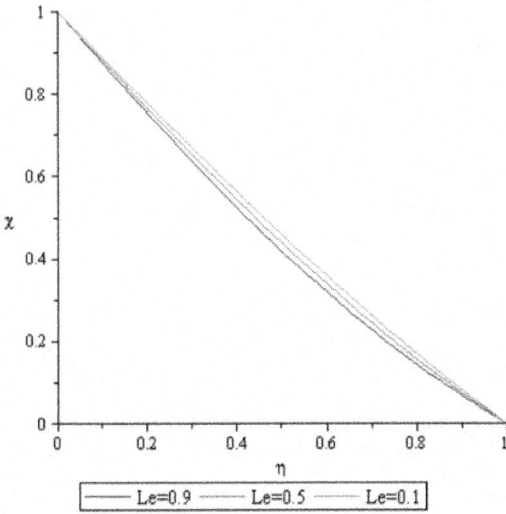

Figure 14: Effect of Le on Nanofraction Profile

Figures 15 to 16 display the effect of Soret number (Sr), and it is observe that, increase in the Soret number causes the temperature and nanofraction boundary thickness to reduce.

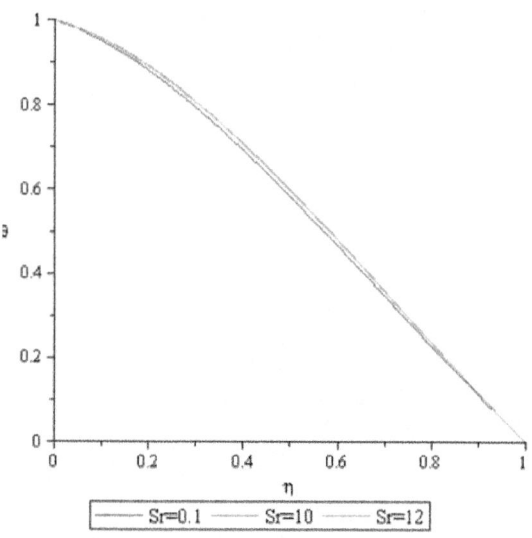

Figure 15: Effect of the Sr on Temperature Profile

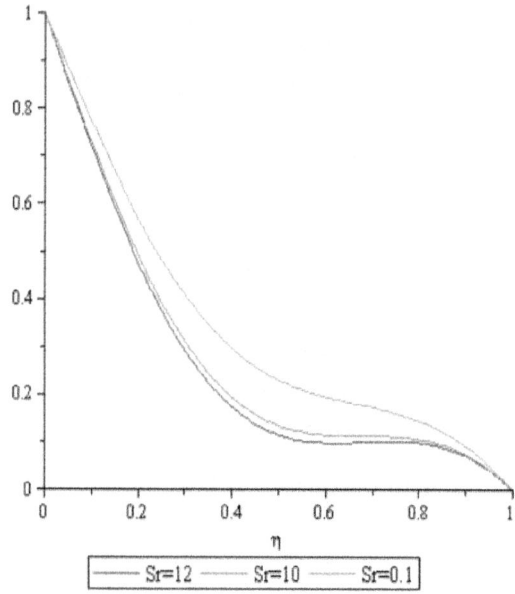

Figure 16: Effect of the Sr on Nanofraction Profile

Figures 17 to 18 shows that Brownian motion (N_b) causes the temperature profile and the nanofraction profile to be enhance.

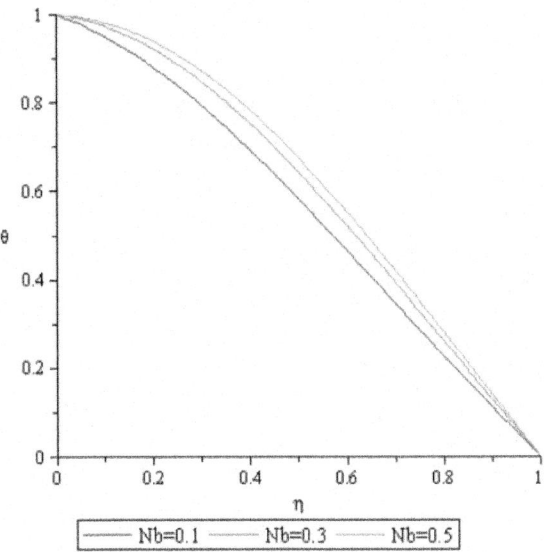

Figure 17: Effect of Nb on Temperature Profile

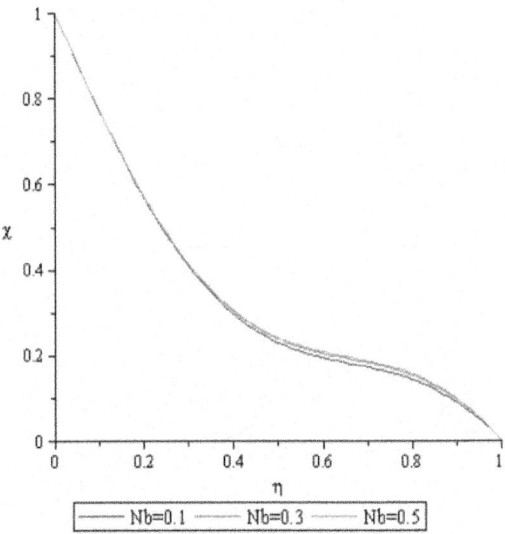

Figure 18: Effect of Nb on Nanofraction Profile

Figures 19 to 20 shows that increase in Thermophoresis parameter (N_t) causes the temperature profile to be enhance while the concentration profile reduces.

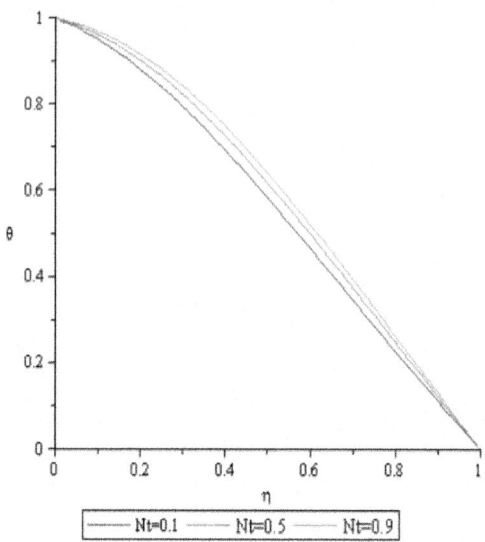

Figure 19: Effect of Nt on Temperature Profile

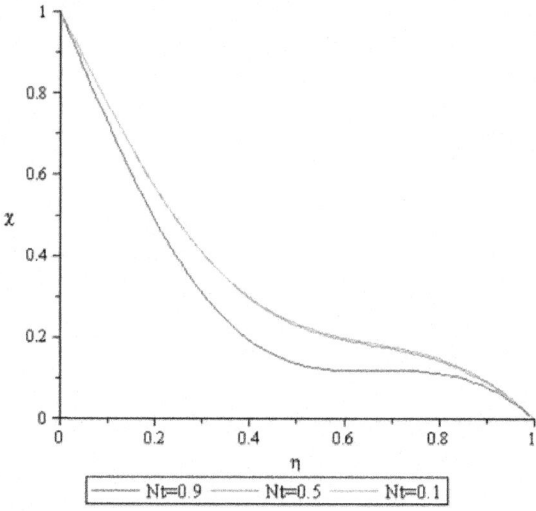

Figure 20: Effect of Nt on Nanofraction Profile

CONCLUSIONS

The solution to the problem of laminar fluid flow in an inclined parallel walls resulting from the movement of the lower wall while the upper wall remain stationary (Coette flow) in a nanofluid with thermal convection, Soret and

Dufour effects with radiation has been obtained using the Modified Adomian Decomposition Method for the first time. The model used for the nanofluid was presented in its rectangular coordinate system and incorporates the effect of Brownian motion, and thermophoresis parameter. A similarity solution was presented which depends on the Prandtl number P_r, Lewis number Le, Brownian motion N_b, thermophoresis number N_t, Soret (Sr) number, Dufour (DU) number, Schimidt number Sc, and Grashof numbers Gr_T, Gr_c. It was found that:-

- All the graphs presented in this work satisfy the boundary conditions.

- It is an established that Soret and Dufour are opposite of one another and this work shows that increase in Soret number leads to reduction in temperature and nanofraction profiles while increase in Dufour number enhances the temperature and nanofracion profiles.

- The results obtained in this work are in good agreement with the Numerical Methods as shown in Table 1 which proves the efficiency of the method.

- Larger values of Prandtl number are equivalent to increase in the thermal conductivity of the fluid and therefore heat is able to diffuse away from the heated surface more rapidly for smaller values of Prandtl number. Hence there is a reduction in temperature with decrease in the Prandtl number.

- It is generally observed from the graphs, that the velocity, temperature and nanofraction profile are at the maximum on the lower channel (moving wall) while they are at minimum on the upper channel (stationary wall). These clearly represent the idea of the authors from the model formulation (realistic).

- It is also observed in Figure 5 that the higher the Nanofraction Grashof number, the higher the velocity of the fluid at the lower channel; But falls quickly to zero as it approaches the upper wall than for smaller values of nanofraction Grashof number.

- It should be noted that as a particular quantity is varied, all others are kept constant.

REFERENCES

1.   E. Abu-Nada, Z. Masoud, A. Hijazi (2008). Natural convection heat transfer enhancement in horizontal concentric annuli using nanofluids, Int. Commun. Heat Mass Transfer 35, 657–665.

2.   R. Ellahi (2013). The effects of MHD and temperature dependent viscosity on the flow of non-Newtonian nanofluid in a pipe: analytical

solutions, Appl. Math. Modell. 37 (3), 1451–1467.

3.   M. Sheikholeslami, M. Gorji-Bandpy, D.D. Ganji (2013). Numerical investigation of MHD effects on Al2O3–water nanofluid flow and heat transfer in a semiannulus enclosure using LBM, Energy 60 501–510.

4.   M.M. Rashidi, S. Abelman, N. Freidooni Mehr (2013). Entropy generation in steady MHD flow due to a rotating porous disk in a nanofluid, Int. J. Heat Mass Transfer 62, 515–525.

5.   M. Sheikholeslami, M. Gorji-Bandpy, S. Soleimani (2013). Two phase simulation of nanofluid flow and heat transfer using heatline analysis, Int. Commun. Heat Mass Transfer 47, 73–81.

6.   M. Sheikholeslami, M. Gorji-Bandpy, R. Ellahi, A. Zeeshan (2014). Simulation of MHD CuO–water nanofluid flow and convective heat transfer considering Lorentz forces, J. Mag. Magn. Mater. 369, 69–80.

7.   A.Sh. Kherbeet, H.A. Mohammed, B.H. Salman (2012). The effect of nanofluids flow on mixed convection heat transfer over microscale backward-facing step, Int. J. Heat Mass Transfer 55, 5870–5881.

8.   Aiyesimi, Y.M., Yusuf, A., Jiya, M. (2015). Hydromagnetic Boudary-Layer Flow of a Nanofluid Past a Stretching Sheet Embedded in a Darcian Porous Medium with Radiation. Nigerian Journal of Mathematics and Applications, 24, 13-29.

9.   Aiyesimi, Y.M. Yusuf, A. & Jiya, M. (2015). An Analytic Investigation Of Convective Boundary-Layer Flow Of A Nanofluid Past A Stretching Sheet With Radiation. Journal of Nigerian Association of Mathematical Physics. 29 (1), 477-490.

10.   Khan, W.A., Pop, I. (2010). Boundary-layer flow of a nanofluid past a stretching sheet, Int. J. Heat Mass Transf. 53, 2477-2483 (2010). S. M. Metev and V. P. Veiko, Laser Assisted Microtechnology, 2nd ed., R. M. Osgood, Jr., Ed. Berlin, Germany: Springer-Verlag, 1998.

11.   Sheikholeslami, M., Abelman, S., Ganji, D. D. (2014). Numerical simulation of MHD nanofluid flow and heat transfer considering viscous dissipation. International Journal of Heat and Mass Transfer 79, 212–222.

12.   Ebaid, A. & Al-Armani, N. (2013). A new Approach for a Class of the Blasius Problem via a Transformation and Adomian's Method. Abstract and Applied Analysis, the Scientific World Journal, 2013.

Chapter 6

CONSTITUTIVE RELATIONS OF STRESS AND STRAIN IN STOCHASTIC FINITE ELEMENT METHOD

Drakos Stefanos

International Centre for Computational Engineering, Rhodes, Greece

ABSTRACT

The analysis and design in structural and geotechnical engineering problems requires the calculation of stress and strain which is generally a difficult task because of the uncertainty and spatial variability of the properties of soil materials. This paper presents a procedure of conducting Stochastic Finite Element Analysis using Polynomial Chaos in order to propagate the uncertainties of input to constitutive relation of stress and strain. The problem is dominated by highly non linearity. Among other methods the procedure leads to an efficient computational cost for real practical problems. This is achieved by polynomial chaos expansion displacement field, stress and strain also. An example of a plane-strain strip load on a semi-infinite elastic foundation is presented and the results of settlement are compared to those obtained from the closed form solution method. A close matching of the two is observed. The constitutive relation of stress and strain is presented as result of the Polynomial Chaos expansion and Monte Carlo method. A close matching of the two method is observed also.

INTRODUCTION

The analysis and design in structural and geotechnical engineering problems requires the calculation of stress and strain which is generally a difficult task because of the uncertainty and spatial variability of the materials's properties. Various forms of uncertainties arise which depend on the nature of geological formation or construction method, the site investigation, the type and the accuracy of design calculations etc. In recent years there has been considerable interest amongst engineers and researchers in the issues related to quantification of uncertainty as it affects safety, design as well as the cost of projects.

A number of approaches using statistical concepts have been proposed in engineering in the past 25 years or so. These include the Stochastic Finite Element Method (SFEM) [1-3], and the Random Finite Element Method (RFEM) [4-8]. The RFEM involves generating a random field of soil or structure properties with controlled mean, standard deviation and spatial correlation length, which is then mapped onto a finite element mesh. However the number of works on the stochastic stress and strain calculation and their statistical moments are limited. An essential paper on the field is presented by Ghosh & Farhat [9] where the constitutive relation of stress and strain calculated by different approaches.

In the past SFEM has been developed using different expansions of stochastic variables. In this paper we present SFEM [11-13] using the method of Generalized Polynomial

Chaos (GPC) [14]. To descretise the stochastic process of material the Karhunen-Loeve Expansion was used and it is presented. The constitutive relation of stress and strain calculated using the Generalized Polynomial Chaos and verified against Monte Carlo simulation which is treated as the exact solution based on a series of computational experiment.

A numerical example of foundation settlement given in the last part of the paper and the results of settlement compared with those arises from closed form solution. The two methods of stress and strains constitutive relation compared also and the results are presented.

PROBLEM DESCRIPTION AND MODEL FORMULATION

Let us consider a general boundary value problem of computation of probable deformation of a body of arbitrary shape having randomly varying material properties caused by the application of a randomly varying load as shown in Fig. 1.

According to the elasticity theory a boundary value problem can be described as follow:

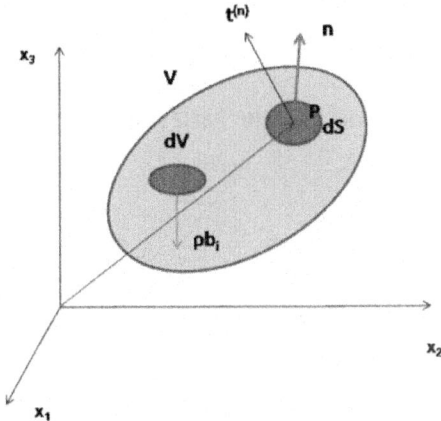

Figure 1: Body of arbitrary shape

$$\begin{cases} \sigma_{ij,j}(x,\omega) = f(x,\omega) \ in \ D \times \Omega \\ \quad \sigma_{ij}(x,\omega) = C_{ijkl}(x,\omega)\varepsilon_{kl}(x,\omega) \ in \ D \times \Omega \\ \quad u(x,\omega) = g_D \ in \ B_D \\ \sigma_{ij}(x,\omega)n_j = g_N \ in \ B_D \end{cases}$$

(1)

And in the weak form as:

$$a(u,v) = l(v)$$

(2)

Where:

$$a(u,v) = \int_D \varepsilon^T(v)C(x,\omega)\varepsilon(u) \, dx$$

(3)

$$l(v) = \int_D f(x,\omega) \cdot v dx + \int_{B_N} g_N \cdot v dx$$
$$- \int_{B_D} \varepsilon^T(v)C(x,\omega)\varepsilon(u) \, dx \cdot g_D$$

(4)

In order to model the problem assuming the sample space $(\Omega,\mathcal{F},\mathbb{P})$ where $\mathcal{F}$ is the σ-algebra and is considered to contain all the information that is available, $\mathbb{P}$ is the probability measure and the spatial domain of the soil or the structure is $D \subset \mathbb{R}^2$. The Elasticity modulus $\{E(x,\omega): \in D \times \Omega\}$ and the external load $\{f(x,\omega): \in D \times \Omega\}$ considered as second order random fields and their functions are determined $E, f: D \times \Omega \rightarrow \mathbb{R} \in V = L^2(\Omega, L^2(D))$ and characterized by specific distribution where in our case as lognormal. Considering as μ_k, σ_k and $v_k = \frac{\sigma_k}{\mu_k}$ the mean value the standard deviation and the coefficient of variation of Elasticity modulus, the lognormal distribution is given [8]:

$$E = \exp(\mu_{lnK} + \sigma_{lnk} Z(\omega))$$

(5)

Where the mean values and the variance of the distribution are equal to:

$$\begin{cases} \sigma_{lnk}^2 = \ln(1 + v_k^2) \\ \mu_{lnk} = \ln(\mu_k) - \frac{1}{2}\sigma_{lnk}^2 \end{cases}$$

(6)

And $\omega \in \Omega, Z \sim N(0,1)$

To separate the deterministic part from the stochastic part of the formulation the Karhunen-Loeve expansion has been used. It is considered as the most efficient method for the discretization of a random field, requiring the smallest number of random variables to represent the field within a given level of accuracy. Based on that the stochastic process of Young modulus over the spatial domain with a known mean value $\bar{E}(x)$ and covariance matrix $Cov(x_1, x_2)$ assuming lognormal distribution is given by:

$$E(x, \xi(\omega)) = \exp(\tilde{E}(x) + \sum_{\kappa}^{\infty} \sqrt{\lambda_\kappa} w_\kappa(x)\xi_\kappa(\omega))$$

(7)

In practice, calculations were carried out over a finite number of summations (for example 1-5) so the approximate stochastic representation is given by the trancuated part of expansion:

$$E(x, \xi(\omega)) = \exp(\tilde{E}(x) + \sum_{\kappa=1}^{K} \sqrt{\lambda_\kappa} w_\kappa \xi_\kappa(\omega))$$

(8)

Where:

λ_κ are the eingenvalues of the covariance function

$w_\kappa(x)$ are the eingenfunctions of the covariance function $Cov(x_1, x_2)$

$x \in D$ and $\omega \in \Omega$

$\xi = [\xi_1, \xi_1, ..., \xi_M]: \Omega \to \Gamma \subset \mathbb{R}^M$

and

$\Gamma = \Gamma_1 \times \Gamma_1 \times ... \times \Gamma_M$

The pairs of eingenvalues and eingenfunctions arised by the equation:

$$\int_D C(x_1, x_2)\varphi_\kappa(x_2) = \lambda_\kappa w_\kappa(x_1)$$

(9)

Using the Karhunen-Loeve expansion the stochastic elasticity tensor is given by:

$$C_{ijkl}(x, y) = E(x)C^*_{ijkl}(x), \quad i, j, k, l = 1,2,3 \tag{10}$$

$C^*_{ijkl}(x)$ is expressed in terms of (deterministic) Poisson's ratio as

To compute the statistical moments of the calculations output we perform a change of variable $y_k := \xi_k(\omega)$ and $y = [y_1, y_2 \cdots, y_M]$ [10]. If the random variables are independent and ρ_i denote the density of ξ_i then the joint density is given by:

$$p(y) = p_1(y_1)p_2(y_2)\cdots p_M(y_M) \tag{12}$$

Drakos & Pande [12, 13] developed a new algorithm of Stochastic Finite method using the Generalized Polynomial Chaos (appendix A) and proved that the problem formulation has the final form:

$$Q_m \otimes K_m = q_0 \otimes (f_{0+}t_{gN}) - Q_m \otimes K_{Bm} \cdot g_d \tag{13}$$

Where:

$$\begin{cases} Q_m = \int_\Gamma p(y)\psi_k(y)\psi_p(y)\exp(\sum_{k=1}^K \sqrt{\lambda_k}\varphi_k y_k)\, dy \\ K_m = \int_D B^T \exp(\tilde{E}(x))C^*(x)B\,dxdy \end{cases} \tag{14}$$

$$\begin{cases} q_0 = \int_\Gamma p(y)\psi_p(y)\psi_1(y)\, dy \\ K_{Bm} = \int_{BD} B^T \exp(\tilde{E}(x))C^*(x)B\,dxdy \\ f_0 = \int_D \varphi^T f(x)\,dx \\ t_{gN} = \int_{BN} \varphi^T \cdot g_N\,ds \end{cases} \tag{15}$$

B is strain displacement matrix.

φ is the hat function.

ψ is the Polynomial Chaos

CONSTITUTIVE RELATIONS OF STRESS AND STRAIN

The calculation of the constitutive relation of stress and strain in the case of stochastic problems is quite complicated especially when the invariant of them are needed where the equations become highly nonlinear. In [9] several numerical integration schemes to evaluate the statistical moments of strains and stresses in a random system is presented. In the current work the propagation of the input uncertainty to the stress and strain relation is modelled by the polynomial Chaos expansion and verified against Monte Carlo simulation which is treated as the exact solution of the problems. The computational implementation of the Monte Carlo Method leads to the random field generation and the requested function $u_k(x)$ gets a new value for each realization. At the end of all simulations the statistical moment are calculated. The expected value and the variance are given by:

$$\begin{cases} \mathbb{E}(u(x)) = \frac{1}{K}\sum_{k=1}^{K} u_k(x) \\ Var(u(x)) = \frac{1}{K-1}\sum_{k=1}^{K}(u_k(x) - \mathbb{E}(u(x))^2 \end{cases}$$

(16)

In an elastostatic problem of homogeneous isotropic body one of the field equations that must be satisfied at all interior points of the body is the Strain-Displacement relations:

$$\varepsilon_{ij}(x,y) = \frac{1}{2}(u(x,y)_{i,j} + u(x,y)_{j,i}) \quad i,j = 1,2,3$$

(17)

Using the displacement polynomial chaos expansion

$$u(x,y) = \sum_{k=1}^{Q} u_k(x)\psi_\kappa(y)$$

(18)

Where: Q and ψ are given in appendix A

The equation 17 leads to:

$$\varepsilon_{ij}(x,y) = \frac{1}{2}\left(\left(\sum_{k=0}^{Q} u_i^{(k)}(x)\psi_\kappa(y)\right)_{,j} + \left(\sum_{k=0}^{Q} u_j^{(k)}(x)\psi_\kappa(y)\right)_{,i}\right) \Longrightarrow$$

$$\varepsilon_{ij}(x,y) = \frac{1}{2}\left(\sum_{k=0}^{Q} u_{i,j}^{(k)}(x)\psi_\kappa(y) + \sum_{k=0}^{Q} u_{j,i}^{(k)}(x)\psi_\kappa(y)\right) = \sum_{k=0}^{Q} \varepsilon_{ij}^{(k)}\psi_\kappa(y)$$

(19)

Expected Value of Strains

According to the polynomial chaos expansion of strains the expected value can be evaluated by the following:

$$\mathbb{E}[\varepsilon_{ij}(x,y)] = \mathbb{E}\left[\frac{1}{2}\left(\sum_{k=0}^{Q} u_{i,j}^{(k)}(x)\psi_\kappa(y) + \sum_{k=0}^{Q} u_{j,i}^{(k)}(x)\psi_\kappa(y)\right)\right]$$

$$= \frac{1}{2}\left(u_{i,j}^{(0)}(x)\underbrace{\mathbb{E}[\psi_0(y)]}_{1} + \sum_{k=1}^{Q} u_{i,j}^{(k)}(x)\underbrace{\mathbb{E}[\psi_\kappa(y)]}_{0} + u_{j,i}^{(0)}(x)\underbrace{\mathbb{E}[\psi_0(y)]}_{1} + \sum_{k=1}^{Q} u_{j,i}^{(k)}(x)\underbrace{\mathbb{E}[\psi_\kappa(y)]}_{0}\right)$$

(20)

$$\Rightarrow \mathbb{E}\big[\varepsilon_{ij}(x,y)\big] = \frac{1}{2}\Big(u_{i,j}^{(0)}(x) + u_{j,i}^{(0)}(x)\Big) = \varepsilon_{ij}^{(0)}$$

Variance of Strains

Respected to the expected value evaluation and the Chaos Polynomial characteristics the variance of the strain tensor can be calculated as:

$$\sigma^2 = \mathbb{E}\big[\varepsilon_{ij}^2(x,y)\big] - (\mathbb{E}\big[\varepsilon_{ij}(x,y)\big])^2$$

$$= \sum_{k=0}^{P}[\varepsilon_{ij}^{(k)}(x)]^2 \int_{\Gamma} \rho(y)\psi_k^2(y)dy - [\varepsilon_{ij}^{(0)}(x)]^2$$

$$= [\varepsilon_{ij}^{(0)}(x)]^2 \underbrace{\int_{\Gamma} \rho(y)\psi_0^2(y)dy}_{1} + \sum_{k=1}^{Q}[\varepsilon_{ij}^{(k)}(x)]^2 \int_{\Gamma} \rho(y)\psi_k^2(y)dy - [\varepsilon_{ij}^{(0)}(x)]^2 \Rightarrow$$

$$\sigma^2 = \sum_{k=1}^{Q}[\varepsilon_{ij}^{(k)}(x)]^2 < \psi_k^2(y) >$$

(21)

Expected Value of Stress Tensor

The constitutive relation of stress and strain given by the well known equation of Hooke's law equation:

$$\sigma_{ij}(x,y) = E(x,y)C_{ijmn}^* \varepsilon_{mn} \tag{22}$$

Using the polynomial chaos expansion of strains we get:

$$\sigma_{ij}(x,y) = E(x,y)C_{ijmn}^* \sum_{k=0}^{Q} \varepsilon_{mn}^{(k)}\psi_k(y) \tag{23}$$

According to the elasticity modulus distribution and the strain tensor Chaos Polynomial expansion the expected value of stress tensor takes the following form:

$$< \sigma_{ij}(x,y) >=< E(x,y)C_{ijmn}^* \varepsilon_{mn} >$$

$$=< \exp(mu_{lnE} + \sigma_{lnE}\,y)\,C_{ijmn}^* \sum_{k=0}^{Q} \varepsilon_{mn}^{(k)}\psi_k(y) >$$

$$= e^{mu\,lnE}\,C_{ijmn}^* \sum_{k=0}^{Q} \varepsilon_{mn}^{(k)} \int_{\Gamma} \rho(y)e^{\sigma_{lnE}\,y}\psi_k(y)dy) \Rightarrow$$

$$< \sigma_{ij}(x,y) >= e^{mu\,lnE}\,C_{ijmn}^* \sum_{k=0}^{Q} \varepsilon_{mn}^{(k)} < e^{\sigma_{lnE}\,y}\psi_k(y) > \tag{24}$$

Variance of Stress Tensor

Having calculated the expected value of stress tensor and knowing its stochastic equation of by the Hook's law constitutive relation, the variance of stress tensor can be calculated as:

$$\sigma^2 = \mathbb{E}[\sigma_{ij}^2(x, y)] - (\mathbb{E}[\sigma_{ij}(x, y)])^2$$

$$=< \left[E(x, y) C_{ijmn}^* \sum_{k=0}^{P} \varepsilon_{mn}^{(k)} \psi_\kappa(y) \right]^2 > - [e^{mu\, lnE}\, C_{ijmn}^* \sum_{k=0}^{P} \varepsilon_{mn}^{(k)} < e^{\sigma\, lnE\, y} \psi_k(y) >]^2$$

Given the elasticity modulus distribution the variance of stress tensor is equal to:

$$\sigma^2 =< \left[\exp(mu_{lnE} + \sigma_{lnE}\, y)\, C_{ijmn}^* \sum_{k=0}^{Q} \varepsilon_{mn}^{(k)} \psi_\kappa(y) \right]^2 > - \left[e^{mu\, lnE}\, C_{ijmn}^* \sum_{k=0}^{Q} \varepsilon_{mn}^{(k)} \int_r \rho(y) e^{\sigma_{lnE}\, y} \psi_k(y) dy \right]^2 \tag{25}$$

Pore Pressure Calculation

A major issue in a wide range of geotechnical and geomechanics problem is the estimation of the excess pore pressure in the ground. Using the Chaos Polynomial expansion the statistical moments of pore pressure can be evaluated as following. The pore pressure is given by the following equation:

$$p = K_a \varepsilon_v = K_a (\varepsilon_{11} + \varepsilon_{22} + \varepsilon_{33})$$

Where:

ε_v is the volumetric strain

is shown [15] as

$$K_a \geq 20 \frac{E'}{(1-2v)} \tag{26}$$

Assuming a minimum value of $K_a = 20 \frac{E'}{(1-2v)}$

Expected Value of Pore Pressure

The expected value of pore pressure is the result of the summation of the expected value of the strain's Chaos expansion multiplied by the stochastic fluid modulus:

$$E[p] = E[K_a(\varepsilon_{11} + \varepsilon_{22} + \varepsilon_{33})] \tag{27}$$

Replacing the longnormal distribution o elasticity modulus we get:

$$E[p] = E\left[20\frac{exp(mu_{lnE} + \sigma_{lnE} y)}{(1-2v)}\left(\sum_{k=1}^{Q}\varepsilon_{11}^{(k)}\psi_\kappa(y) + \sum_{k=1}^{P}\varepsilon_{22}^{(k)}\psi_\kappa(y) + \sum_{k=1}^{Q}\varepsilon_{33}^{(k)}\psi_\kappa(y)\right)\right]$$

$$= 20\frac{exp(mu_{lnE})}{(1-2v)}E\left[e^{\sigma_{lnE} y}\left(\sum_{k=1}^{Q}\varepsilon_{11}^{(k)}\psi_\kappa(y) + \sum_{k=1}^{Q}\varepsilon_{22}^{(k)}\psi_\kappa(y) + \sum_{k=1}^{Q}\varepsilon_{33}^{(k)}\psi_\kappa(y)\right)\right]$$

And finally

$$E[p] = 20\frac{exp(mu_{lnE})}{(1-2v)}\sum_{k=1}^{Q}[\varepsilon_v^{(k)}]\int_\Gamma e^{\sigma_{lnE} y}\psi_\kappa(y)dy \tag{28}$$

Variance of Pore Pressure

Using the outcome of the pore pressure expected value its variance is given by:
$$Var[p] = E[p^2] - (E[p])^2$$

Where:

$$\begin{cases} E[p^2] = E\left[\left(20\frac{exp(mu_{lnE})}{(1-2v)}e^{\sigma y}(\sum_{k=1}^{Q}\varepsilon_v^{(k)}\psi_\kappa(y))\right)^2\right] \\ (E[p])^2 = \left(20\frac{exp(mu_{lnE})}{(1-2v)}\sum_{k=1}^{Q}[\varepsilon_v^{(k)}]\int_\Gamma e^{\sigma_{lnE} y}\psi_\kappa(y)dy\right) \end{cases} \tag{29}$$

Invariants of Stress Tensor

It is well known that to solve engineering problem the model itself should be defined independent of the coordinate system attached to the material. Thus, it is necessary to define the model in terms of stress invariants which are, by definition, independent of the coordinate system selected. The physical content of a stress tensor is reflected exclusively in the stress invariants (I_1, I_2, I_3).

Where:

$$I_1 = \frac{1}{3}\sigma_{kk} \tag{30}$$

$$I_2 = \frac{1}{2}\left((\sigma_{kk})^2 - \sigma_{ij}\sigma_{ij}\right) \tag{31}$$

$$I_3 = \varepsilon_{ijk}\sigma_{i1}\sigma_{j2}\sigma_{k3} \tag{32}$$

Solving a stochastic problem the statistical moments of stress invariants are needed. In the following paragraphs the calculation of the expected and variance value of each of invariants are presented.

Expected value of I_1

According to the linearity of the excepted value of I_1 can be calculated:

$$E[I_1] = E[\frac{1}{3}\sigma_{qq}] \Rightarrow$$

$$E[I_1] = \frac{1}{3}(E[\sigma_{qq}])$$

Where:

$$E[\sigma_{qq}] = \Sigma_{k=1}^Q (C_{qqmn} \, \varepsilon_{mn}^{(k)}) \int_\Gamma \rho(y) e^{\sigma \ln E \, y} \psi_\kappa(y) dy \tag{33}$$

And

$$C = \exp(mu_{lnE}) \cdot C^* \tag{34}$$

Variance of I_1

Knowing the mean value of I_1 the variance is:

$$Var(I_1) = E[I_1^2] - (E[I_1])^2$$

But due to linearity:

$$Var(I_1) = \frac{1}{9}(Var[\sigma_{qq}])$$

Where:

$$\begin{cases} E[I_1^2] = \frac{1}{9}E[\sigma_{qq}^2] = E\left[\left(e^{\sigma \ln E \, y} C_{qqmn} \, \Sigma_{k=1}^Q \, \varepsilon_{mn}^{(k)} \psi_\kappa(y)\right)^2\right] \\ (E[I_1])^2 = \frac{1}{9}(E[\sigma_{qq}])^2 = \left(\Sigma_{k=1}^Q (C_{qqmn} \, \varepsilon_{mn}^{(k)}) \int_\Gamma \rho(y) e^{\sigma y} \psi_\kappa(y) dy\right)^2 \end{cases} \tag{35}$$

Expected Value of I_2

As shown before:

$$E[I_2] = E[\frac{1}{2}((\sigma_{qq})^2 - \sigma_{ij}\sigma_{ij})]$$

This gives

$$E[I_2] = \frac{1}{2}\left(E\left[\left(e^{\sigma \ln E}{}^y C_{qqmn} \sum_{k=1}^{Q} \varepsilon_{mn}^{(k)} \psi_\kappa(y) \right)^2 \right] - E\left[e^{\sigma \ln E}{}^y C_{ijmn} \sum_{k=1}^{Q} \varepsilon_{mn}^{(k)} \psi_\kappa(y) \cdot e^{\sigma \ln E}{}^y C_{ijmn} \sum_{k=1}^{Q} \varepsilon_{mn}^{(k)} \psi_\kappa(y) \right] \right)$$

(36)

Variance of I_2

The variance of I_2 due to the stress product on its equation become highly nonlinear. Thus

$$Var(I_2) = E[I_2^2] - (E[I_2])^2$$

Where:

$$\begin{cases} E[I_2^2] = \frac{1}{4} E\left[\begin{array}{c} \left[\left(e^{\sigma \ln E}{}^y C_{qqmn} \sum_{k=1}^{Q} \varepsilon_{mn}^{(k)} \psi_\kappa(y) \right)^2 \right] \\ -\left(E\left[e^{\sigma \ln E}{}^y C_{ijmn} \sum_{k=1}^{Q} \varepsilon_{mn}^{(k)} \psi_\kappa(y) \cdot e^{\sigma \ln E}{}^y C_{ijmn} \sum_{k=1}^{Q} \varepsilon_{mn}^{(k)} \psi_\kappa(y) \right] \right)^2 \end{array} \right] \\ (E[I_2])^2 = \left(\left[\frac{1}{2} \left(\begin{array}{c} E\left[\left(e^{\sigma \ln E}{}^y C_{qqmn} \sum_{k=1}^{Q} \varepsilon_{mn}^{(k)} \psi_\kappa(y) \right)^2 \right] - \\ E\left[e^{\sigma \ln E}{}^y C_{ijmn} \sum_{k=1}^{Q} \varepsilon_{mn}^{(k)} \psi_\kappa(y) \cdot e^{\sigma \ln E}{}^y C_{ijmn} \sum_{k=1}^{Q} \varepsilon_{mn}^{(k)} \psi_\kappa(y) \right] \end{array} \right) \right] \right) \end{cases}$$

(37)

Expected Value of I_3

The high non linearity presented also in the statistical moments of I_3.

$$E[I_3] = E[e_{ijk} \sigma_{i1} \sigma_{j2} \sigma_{k3}]$$

This leads to:

$$E[I_3] = e_{ijk} E\left[e^{3\sigma \ln E}{}^y C_{i1mn} \sum_{k=1}^{Q} \varepsilon_{mn}^{(k)} \psi_\kappa(y) \cdot C_{j2mn} \sum_{k=1}^{Q} \varepsilon_{mn}^{(k)} \psi_\kappa(y) \cdot C_{k3mn} \sum_{k=1}^{Q} \varepsilon_{mn}^{(k)} \psi_\kappa(y) \right]$$

(38)

Variance of I_3

Similarly as documented above the variance of I_3 is equal to:

$$Var(I_3) = E[I_3^2] - (E[I_3])^2$$

Where:

$$E\left[\left(e_{ijk} e^{3\sigma \ln E}{}^y C_{i1mn} \sum_{k=1}^{Q} \varepsilon_{mn}^{(k)} \psi_\kappa(y) \cdot C_{j2mn} \sum_{k=1}^{Q} \varepsilon_{mn}^{(k)} \psi_\kappa(y) \cdot C_{k3mn} \sum_{k=1}^{Q} \varepsilon_{mn}^{(k)} \psi_\kappa(y) \right)^2 \right] - $$

$$\left(E\left[e_{ijk} e^{3\sigma \ln E}{}^y C_{i1mn} \sum_{k=1}^{Q} \varepsilon_{mn}^{(k)} \psi_\kappa(y) \cdot C_{j2mn} \sum_{k=1}^{Q} \varepsilon_{mn}^{(k)} \psi_\kappa(y) \cdot C_{k3mn} \sum_{k=1}^{Q} \varepsilon_{mn}^{(k)} \psi_\kappa(y) \right] \right)^2$$

(39)

NUMERICAL EXAMPLE

A shallow foundation problem for various values of variation's coefficient v_e is solved taken to account the randomness of the ground. To estimate the statistical moments of the soil deformation the numerical algorithm of SFEM using the Generalized Polynomial Chaos as described in the previous paragraphs is applied and the results are compared to those obtained by the closed form solution. To avoid the negative values of the elastic modulus assumed to have lognormal. It is known that the settlement beneath a foundation with uniform but random elastic modulus is given by the equation [8]:

$$S = \frac{S_{det}\, \mu_E}{E}$$

(40)

Where: S_{det} is the deterministic value of settlement with $E = \mu_E$ everywhere.

Assuming lognormal distribution for the settlements the mean values is equal to

$$\mu_{ln(s)} = ln(s_{det}) + \frac{1}{2}\sigma_{ln(E)}^2$$

(41)

The geometry of the finite elements used for the simulation of the problem presented in Fig. 2. The input data of the problem is the random field modulus with a constant average value equal to 100 Mpa and a fixed Poisson ratio equal to 0.25. Calculations have been made for ten different coefficients $v_e = \frac{\sigma_E}{\mu_E}$ of the elastic modulus with a minimum value of 0.1 and then with step 0.1 to a maximum value equal to 1. The randomness of Elasticity modulus in Fig. 3 is shown. For SFEM one dimensional Hermite GPC with order 5 [14] were used. In the Fig. B1 the results of SFEM method comparatively with the closed form solution are shown and they present great accuracy. It is observed that for of $v_e = 0.5$ the error is equal to 0.8%. In the figures B2-B13 the strains and stress components, the pore pressure and the stress tensor invariants are presented as resulted by the Chaos Polynomial expansion and compared with those raised by the Monte Carlo Method. Simulations of 1000-5000 samples were carried and the convergence of the outcomes decreases as the number of Monte Carlo simulations increases.

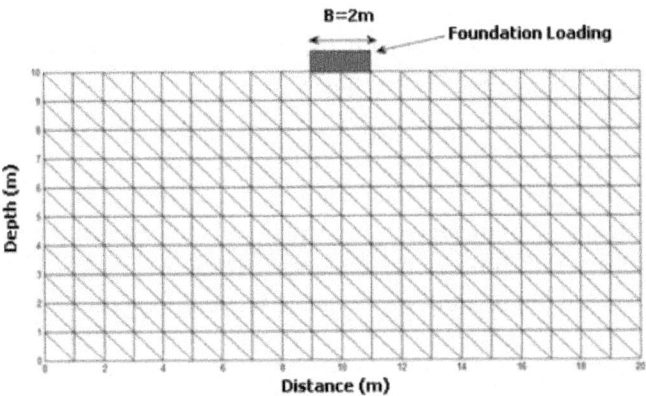

Figure 2: Finite element mesh

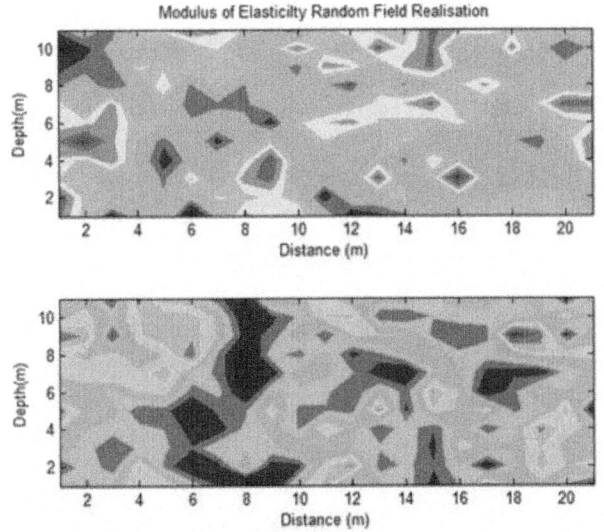

Figure 3: Modulus of Random Elasticity of two different realizations

CONCLUSIONS

To propagate the uncertainties of input parameters to constitutive relations of strain and stress where arises due to spatial variability of mechanical parameters of soil/rock in geotechnical and geomechanics problems, a procedure of conducting a Stochastic Finite Element Analysis has been presented.

An algorithm of Stochastic Finite Element using Polynomial Chaos has been developed. An analysis of settlement of a plane strain strip load on an elastic foundation has been given as an example of the proposed approach. It is shown that the results of SFEM using polynomial chaos compare well with those obtained from closed form solution.

The stress and strain constitutive relation the pore pressure and the stress invariants are modeled by the polynomial Chaos expansion and verified against Monte Carlo simulation which is treated as the exact solution of the problems. The main advantage in using the proposed methodology is that a large number of realizations which have to be made for (Random Finite Element Method) avoided, thus making the procedure viable for realistic practical problems.

APPENDIX A

Galerkin approximation and Generalized Polynomial of chaos

In order to solve the problem 1 we have to create the new space $L^2_p(\Gamma, H^1_0(D))$ For that reason the subspace $S^k \subset L^2_p(\Gamma)$ is considered as [10].

$$S^k = span\{\psi_1, \psi_2, ..., \psi_\kappa\} \tag{A1}$$

Using the dyadic product of the space V^h, S^k the space $L^2_p(\Gamma, H^1_0(D))$ created. Thus

$$V^{hk} = V^h \otimes V^k = span\{\varphi_i \psi_j, i = 1 ... N, j = 1, ... Q\} \tag{A2}$$

The space V^{hk} has dimension QN and regards the test function v. In the case where exists N_B finite element supported by boundaries condition then the subspace of solution belongs is:

$$W^{hk} = V^{hk} \oplus span\{\varphi_{N+1}, \varphi_{N+2}, ..., \varphi_{N+NB}\} \tag{A3}$$

Assuming that the S^k_i represents a space of univariate orthonormal polynomial of variable $y_i \subset \Gamma_i \subset \mathbb{R}$ with order k or lower and:

$$S^k_i = span\{P^i_{a_i}(y_i), a_i = 0,1,2, ... k\}, i = 1, ... M \tag{A4}$$

The tensor product of the M S^k_i subspace results the space of the Generalized Polynomial Chaos:

$$S^k = S_1 \otimes S_2 ... \otimes S_M \tag{A5}$$

And using (A4)

$$S^k = span \left\{ \prod_{i=1}^{M} \begin{array}{l} P_{a_i}^i(y_i): a_i = 0,1, \cdots k, i = 1 \cdots M, \\ |a| \le k \end{array} \right\}$$

(A6)

Where $|a| = \sum_{i=1}^{M} a_i$

And

$$Q = \dim(S^k) = \frac{(M+k)!}{M!k!}$$

(A7)

Xiu & Karniadakis [14] show the application of the method for different kind of orthonormal polynomials and in the current paper the Hermite polynomial was used with the following characteristics:

$$P_0 = 1, < P_i >= 0, i > 0$$
$$< P_m P_n >= \int_\Gamma P_m(y) P_n(y) \rho(y) dy = \gamma_n \delta_{mn}$$

(A8)

Where:

$\gamma_n =< P_n^2 >$ are the normalization factors, δ_{mn} is the Kronecker delta

$\rho(y) = \frac{1}{\sqrt{2\pi}} e^{-\frac{y}{2}}$ is the density function and

$$P_n = (-1)^n e^{\frac{y}{2}} \frac{d^n}{dy^n} e^{-\frac{y}{2}}$$

(A9)

For a 3rd order of one dimension of uncertainty the Hermite Polynomial Chaos is given by:

$$\psi_o(y) = P_0(y) = 1, \quad \psi_1(y) = P_1(y) = y,$$
$$\psi_2(y) = P_2(y) = y^2 - 1$$

APPENDIX B

Results of Numerical Example

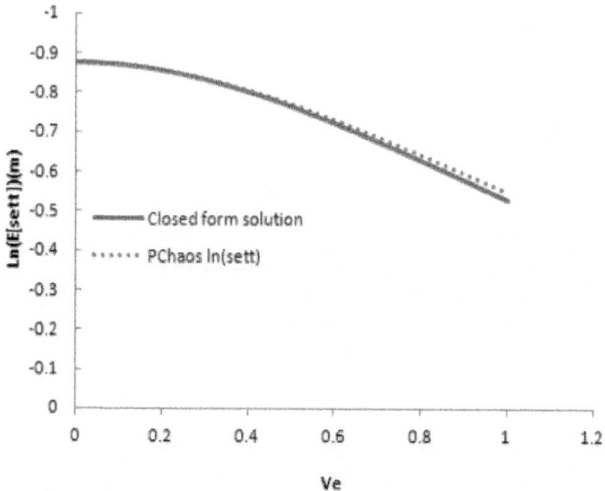

Figure B1: Closed form solution and SFEM results

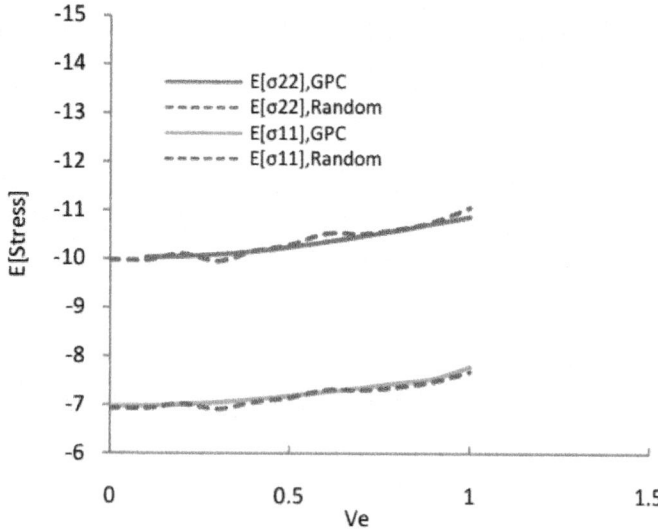

Figure B2: Expected value of stress tensor complements (MC 1000 samples)

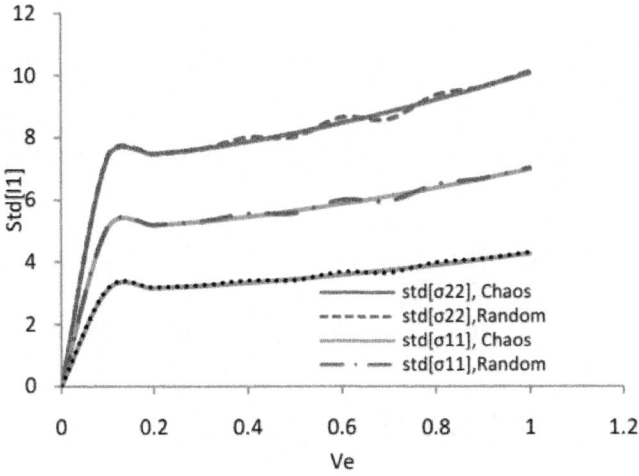

Figure B3: Standard deviation of stress tensor complements. (MC 1000 samples)

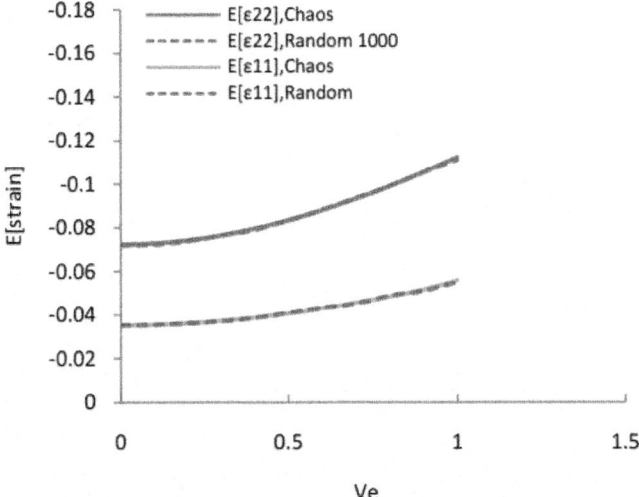

Figure B4: Expected value of strain tensor complements. (MC 1000 samples)

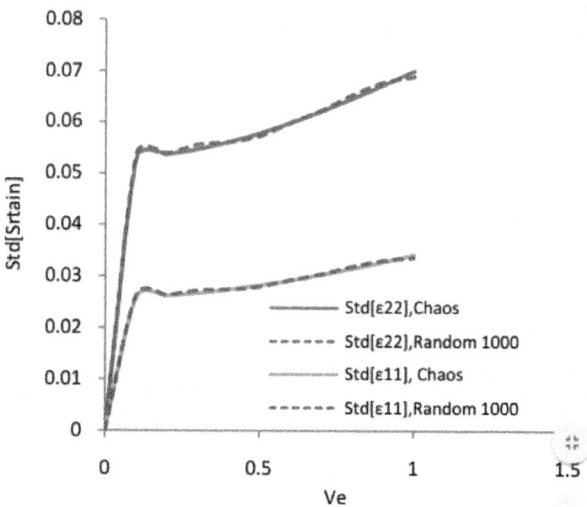

Figure B5: Standard deviation of strain tensor complements. (MC 1000 samples)

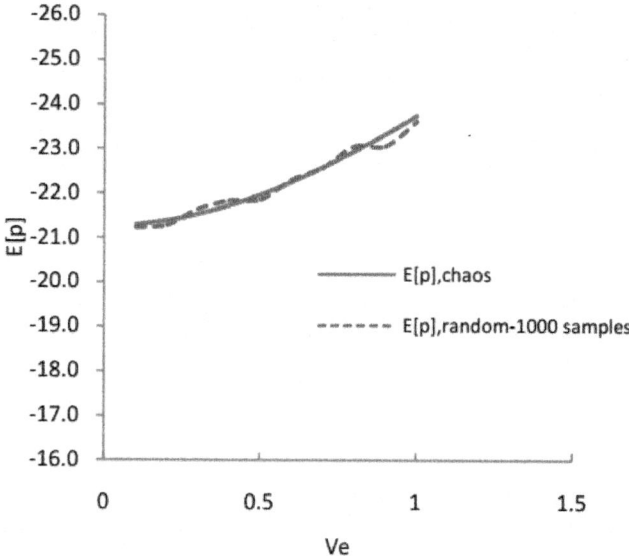

Figure B6: Expected value of pore pressure. (MC 1000 samples)

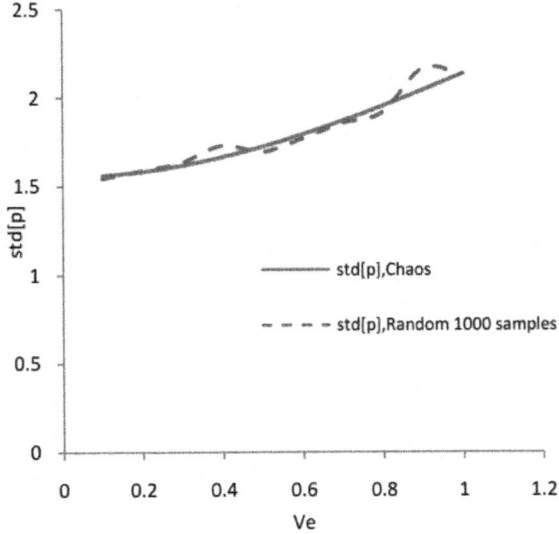

Figure B7: Standard deviation of pore pressure. (MC 1000 samples)

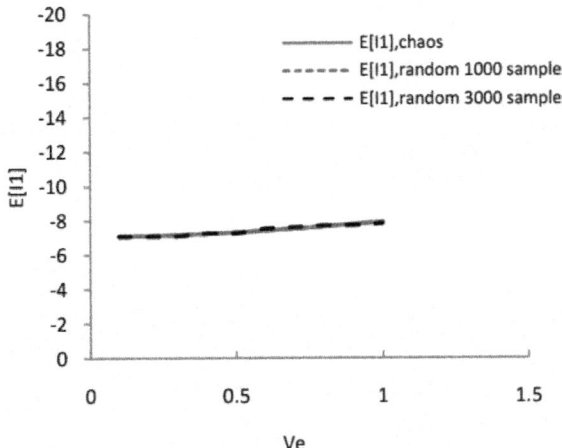

Figure B8: Expected value of stress tensor invariant $II1$. (MC 1000, 3000 samples)

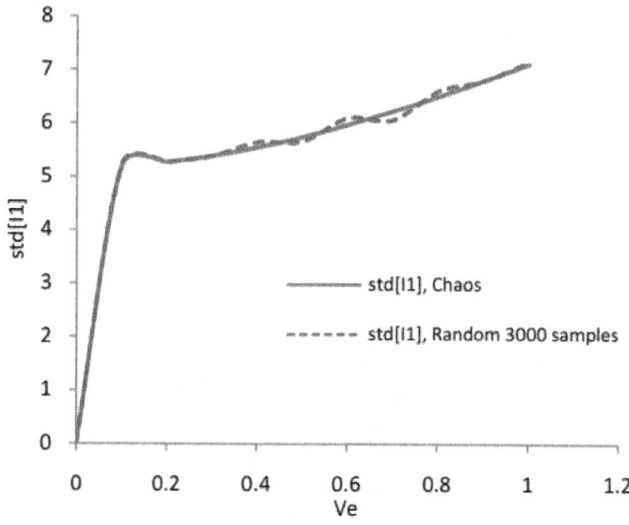

Figure B9: Standard deviation of stress tensor invariant $II1$. (MC 3000 samples)

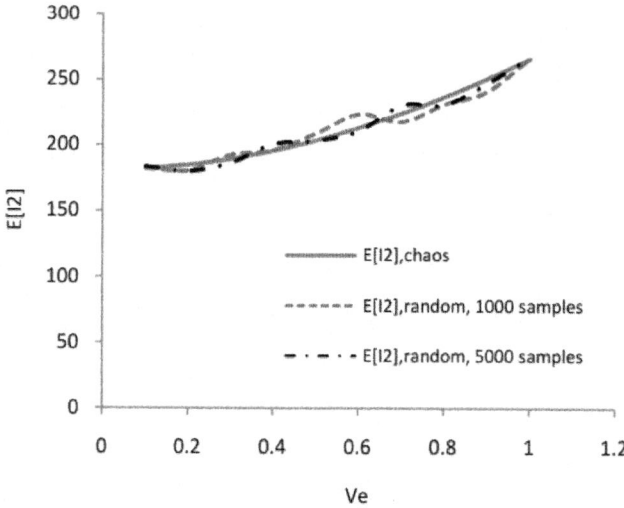

Figure B10: Expected value of stress tensor invariant $II2$. (MC 1000, 5000 samples)

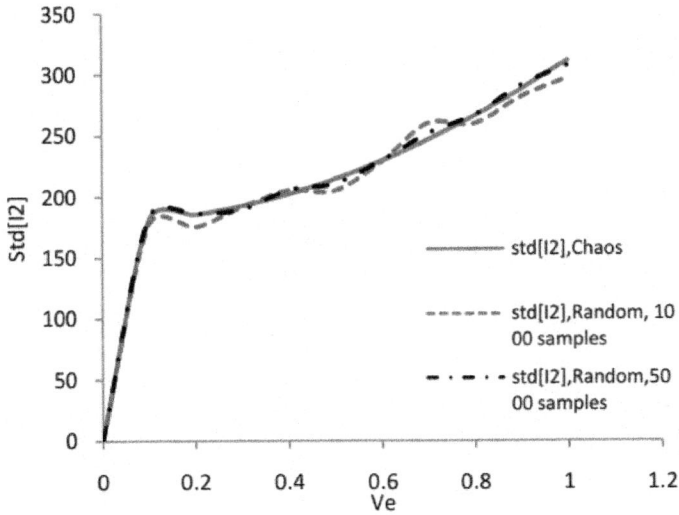

Figure B11: Standard deviation of stress tensor invariant $II2$. (MC 1000, 5000 samples)

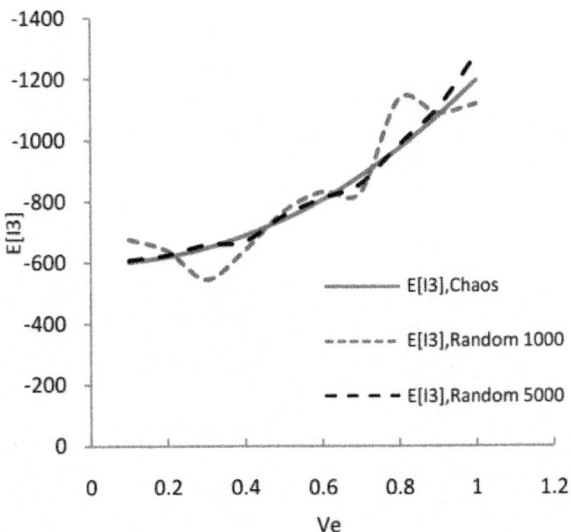

Figure B12: Expected value of stress tensor invariant $II3$. (MC 1000, 5000 samples)

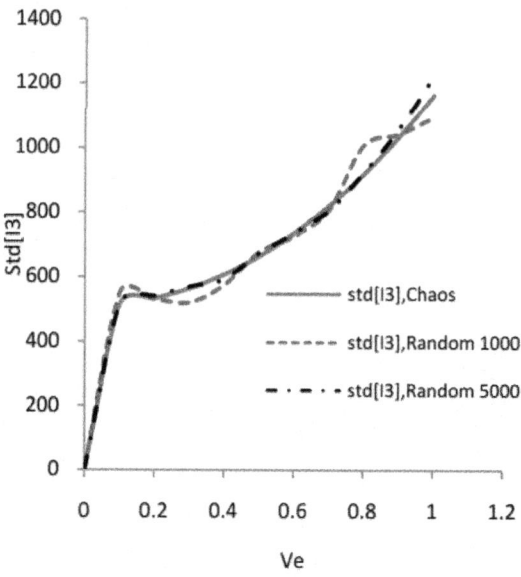

Figure B13: Standard deviation of stress tensor invariant *II*3. (MC 1000, 5000 samples)

REFERENCES

1. Phoon, K., Quek, S., Chow, Y. and Lee, S (1990), "Reliability analysis of pile settlements". Journal of Geotechnical Engineering, ASCE, 116(11), 1717–35.

2. Mellah, R., Auvinet, G. and Masrouri, F. (2000), "Stochastic finite element method applied to non-linear analysis of embankments". Probabilistic Engineering Mechanics, 15 (3), 251–259.

3. Eloseily, K., Ayyub, B. and Patev, R. (2002), "Reliability assessment of pile groups in sands". Journal of Structural Engineering, ASCE, 128(10), 1346–53.

4. Fenton, G. A., and Vanmarcke, E. H. (1990), "Simulation of random fields via local average subdivision." J. Eng. Mech., 116(8), 1733–1749.

5. Paice, G. M., Griffiths, D. V., and Fenton, G. A. (1996), "Finite element modeling of settlements on spatially random soil." J. Geotech. Eng., 122(9), 777–779.

6. Fenton, G. A., and Griffiths, D. V. (2002), "Probabilistic foundation settlement on spatially random soil." J. Geotech. Geoenviron. Eng.,128(5), 381–390.

7. Fenton, G. A., and Griffiths, D. V. (2005), "Three-dimensional probabilistic foundation settlement." J. Geotech. Geoenviron. Eng., 131(2), 232–239.

8. Fenton, G. A., and Griffiths, D. V. (2008), "Risk assessment in geotechnical engineering", Wiley, Hoboken, NJ.

9. Ghosh Debraj and Farhat Charbel (2008), "Strain and stress computations in stochastic finite element methods", International Journal for Numerical Method In Engineering", Vol. 74, Iss. 8, 1219–1239.

10. Lord G., Powel C., Shardlow T. (2014), "An Introduction to Computational Stochastic PDEs", Cambridge Texts in Applied Mathematics.

11. Ghanem, R. G., and Spanos, P. D. (1991), Stochastic finite elements: A spectral approach, Springer-Verlag, New York.

12. Drakos I. S. & Pande G. N. "Quantitative of uncertainties in Earth Structures", World Congress on Advances in Structural Engineering and Mechanics, Incheon Korea 25-29 August, 2015.

13. Drakos I. S. & Pande G. N. "Stochastic Finite Element Analysis using Polynomial Chaos" Studia Geotechnica et Mechanica Journal.(Under Review).

14. D Xiu, G Em Karniadakis. (2003), "Modeling uncertainty in steady state diffusion problems via generalized polynomial chaos" Computer Methods in Applied Mechanics and Engineering 191 (43), 4927-4948.

15. Griffiths, D.V (1985). "The effect of pore-fluid compressibility on failure loads in elasto-plastic soil." Int J Numer Anal Methods Geomech, vol.9, no.3, pp.253-259.

Chapter 7

GENERALIZED GAUSSIAN ELIMINATION (GGE) SOLVING SYSTEM OF LINEAR INEQUALITIES OR EQUALITIES (LIS-II)

Paul T. R. Wang

Wang Paul_Research, Potomac, Maryland, USA

ABSTRACT

The author generalizes the traditional Gaussian Elimination (GE) technique to resolve the feasibility of any system of linear inequalities or equalities. Any linear system consists of either equalities and/or inequalities or mixed of both equalities and inequalities is converted to its homogeneous linear feasibility standard form (HLSF). Variable substitution (VS) in the original Gaussian Elimination is replaced by variable transition (VT) to eliminate a specific variable of choice in a recursive fashion such that at the end only one single variable left to yield the feasible interval for the selected variable if such an interval exists. Note that the feasible interval of a specific variable can be null, single value, bounded below and above, bounded below only, bounded above only, or both unbounded above and below. Furthermore, the feasible interval of a given variable if it exists must also include its integer or binary solution or solutions. It is further shown that the original GE is indeed a special case of the GGE and both GE and GGE share Identical computational complexity that is bounded by the worst case of $O[n^2 m]$. GGE is applicable to any linear system with a finite number of variables, n, and m, a finite number of equalities, inequalities, or mixed constraints. GGE can be used to resolve the feasibility of a given linear system with number of variables and constraints over millions or more. The validity of GGE in dealing very large linear system not only addresses the feasibility of linear systems, it may also resolve the computational complexity of the class of NP-complete (NPC) mystery. This innovative GGE technique is applied to various linear programs with unique solution, unbounded solution, or no solution to illustrate its correctness and applicability. GGE is also shown to be applicable in resolving differential variational inequalities (DVI) for both scientific and engineering applications.

INTRODUCTION

Very large system of linear inequalities with thousands or millions of variables and/or constraints are very tough to resolve. Typically, linear inequalities are solved with the famous Simplex method [1-4] the ellipsoid method [6], or the interior points method [7, 11]. Whether or not a linear system is solvable or whether or not a feasible linear system contains integer or binary solution is one of the most challenging questions known as NP-complete (NPC) for applications in operstions research (OR) [10]. In this paper, the author proposes an innovative approach that generalizes the well known Gaussian Elimination (GE) for linear equalities to resolve the feasibility problem of a system of linear inequalities. Instead of variable subsittution, it is replaced by variables transition to eliminate recursively variables of choice until only one single variable is left to determine the feasible intervl associated of the specific variable. The technique is coined as the Generalized Gaussian Elimination (GGE) for linear systems. Furthermore, it is shown that GE is simply a special case of GGE and that both GE and GGE share the same computational complexity. Simple LPs are used to illustrate for all possible cases of feasibble intervals such as single value (unique), infinite many solutions with distinct upper and lower bounds, or both bounded and unbounded cases.

HOMOGENEOUS LINEAR SYSTEMS FEASIBILITY (HLSF) WITH NEW NOTATIONS [17-19]

Given any system of linear inequalities or equalities in vector and matrix form,

$A\vec{x} \geq \vec{b}$ where A is an n by m matrix, $\vec{x}$ and $\vec{b}$ are m by 1 and n by 1 column vectors respectively.

We define the homogeneous linear system feasibility (HLSF) for $A\vec{x} \geq b$ as follows

$$\vec{f} = A\vec{x} - \vec{b} = (A, -\vec{b}) * \begin{bmatrix} \vec{x} \\ 1 \end{bmatrix} = L * \vec{i} \geq \vec{0} \quad \text{where}$$

$$L = (A, -\vec{b}) \text{ and } \vec{i} = \begin{bmatrix} \vec{x} \\ 1 \end{bmatrix}$$

(1)

The vector $\vec{f} = L * \vec{i} \geq 0$ is referred to as the feasibility vector of the original linear system of inequalities $A\vec{x} = \vec{b}$.

Note that any system or subsystem of linear equalities $A\vec{x}=\vec{b}$ can always be converted into a system or subsystem of linear inequalities as $f = Li = \begin{bmatrix} A & -b \\ -A & b \end{bmatrix} i \geq 0$. The author adopted the following new notations to simplify the illustration of GGE:

Let $v = (v_1, v_2, ---, v_{i-1}, v_i, v_{i+1}, ---, v_n)$ be an 1 by n row vector, and let $\overline{v_{-i}} = (v_1, v_2, ---, v_{i-1}, v_{i+1}, ----, v_n)$ be the 1 by n-1 sub-vector of $\vec{v}$.

With such a new notation, it is clear that for the dot product of two vectors $\vec{v}$ and $\vec{t}$

we have,

$$\vec{v} \bullet \vec{i} = \sum_{k=1}^{n} v_k t_k = v_i t_i + \sum_{k \neq i}^{n} v_k t_k = v_i t_i + \overline{v_{-i}} \bullet \overline{t_{-i}}$$

(2)

For a linear inequality converted to dot product $\vec{v} \bullet \vec{t} \geq c$, we have $v_i t_i + \overline{v_{-i}} \bullet \overline{t_{-i}} \geq c$

Given the following homogenous linear inequality (HLI):

$$L = (A, -\vec{b}) \quad \text{and} \quad \vec{i} = \begin{bmatrix} \vec{x} \\ 1 \end{bmatrix}$$

(3)

Let $L = \begin{pmatrix} L_1 \\ L_2 \\ \vdots \\ L_m \end{pmatrix} = (l_{ij})_{m \times n}$ where

$$L_j = (l_{j1}, l_{j2}, -----, l_{jn})$$

(4)

We also adopt the notation, $f_{-i}(\overrightarrow{x_{-i}}) \geq 0$ to highlight the nonexistence or absence of the variable, x_i, from the function (either an equality or an inequality) for $f(\overrightarrow{x_{-i}}) \geq 0$.

A Literature Overview and Motivation

Traditionally, linear inequalities are resolved as linear programs using either the Simplex methods (Dantzig, 1947), Crisis Cross algorithms (Fukuda & Terlaky, 1997), interior point method (Khachiyan, 1979), projective algorithms (Strang, 1987), path-following algorithms (Gondzio & Terlaky, 1996), and penalty or barrier functions (Nocedal and Wright, 1999). Most approaches centered on iterative searching for feasible points within the n dimensional polytope, i.e., the n-polytope, defined by the constraint linear inequalities. The author with his colleagues [18] proposed a new approach that recursively reduces the worst infeasibility, the sum of all infeasibility, and the number of constraints with the worst infeasibility based upon nonzero coefficients or a subset of the nonzero coefficients that defines the given system of linear inequalities [18]. Such an approach is capable of finding the exact solution if such a feasible solution is unique, a feasible solution if there are more than one solution. For linear system that does not have a feasible solution, this approach is capable of minimizing the sum of all infeasibility or the worst infeasibility, and pinpointing to the relevant coefficients to reveal true conflicts of the linear system [18].

Over a five years period from 2009 to 2014 the author presented three new techniques in dealing with linear systems with both equalities and inequalities were propose by the author and his colleagues during the past few years. The first technique (LIS-I) that examines the atomic components of system of linear inequalities is a set of algorithms that recursively reduce the sum of all infeasibility, maximum infeasibility, and the number of constraints with the maximum infeasibility [17, 18], This paper details a generalized Gaussian Elimination (GGE) to obtain the feasible interval of individual variable as LIS-II. A third technique as LS-III utilizes both projective and orthogonal geometry of unit vectors over the surface an n-dimensional hyperspace as the unit-shell and the concepts of equal distanced points to selected set of points on the unit shell with increasing rank recursively to locate a solution if such a solution does exists [19].

THE GENERALIZED GAUSSIAN ELIMINATION (GGE) ALGORITHM FOR HLSF

Let a HLSF in its standard form as (1) $\vec{f} = A\vec{x} - \vec{b} = (A, -\vec{b}) * \begin{bmatrix} \vec{x} \\ 1 \end{bmatrix} = L * \vec{t} \geq \vec{0}$ where the constraint matrix, Let the HLSF $\vec{f} = L\vec{t} \geq 0$ with n constraints and m variables , then L, is an n by $m+1$ matrix with n linear constraints and variables with $\vec{t} = (x_1 x_2, ..., x_m, 1)^T$. We have the i-th column of L, C_i, and the j-th row of L, R_j is

$$C_i = \begin{bmatrix} l_{1i} \\ l_{2i} \\ : \\ l_{(n-1)i} \\ l_{ni} \end{bmatrix}$$

respectively: and

$$R_j = \begin{bmatrix} l_{j1} & l_{j2} & \cdots & l_{j(m-1)} & l_{jm} & l_{j(m+1)} \end{bmatrix}$$

To eliminate a specific variable, x_i from $f = \vec{L}\vec{i} \geq 0$, GGE requires the following steps:

Step 1: normalization;

Step 2: rows permutation by sign;

Step 3: sorting of rows by decreasing number of nonzero coefficients;

Step 4: replacing rows using binding transition;

Steps 1 to 4 are repeatedly applied to eliminate other variables until only the desirable variable left.

When only one variable is left, the final HLSF uniquely defines both an upper bound and a lower bound as the feasible interval for remaining variable that is not eliminated.

Note that the feasible interval $F_i = [\alpha_i, \beta_i]$ with $\beta_i \leq \alpha_i$ may have the following 6 possible cases:

(i) no solution such that $F_i = [\alpha_i, \beta_i] = \phi$ with $\beta_i < \alpha_i$

(ii) unique solution with $F_i = [\alpha_i, \beta_i] = [r, r]$, i.e., we have $x_i = r$

(iii) bounded interval with infinite many solutions as $F_i = [\alpha_i, \beta_i]$ such that $x_i \in F_i$ with $-\infty < \alpha_i < \beta_i < \infty$

(iv) bounded above as $F_i = [\alpha_i, \beta_i]$ such that $-\infty = \alpha_i < \beta_i < \infty$

(v) bounded below as $F_i = [\alpha_i, \beta_i]$ such that $-\infty < \alpha_i < \beta_i = \infty$

(vi) unbounded above and below as $F_i = [\alpha_i, \beta_i]$ such that $-\infty = \alpha_i < \beta_i = \infty$

Note: (vi) may not exist; it is listed for completeness.

Assume that a selected column variable, x_i, is to be eliminated, the detailed description of each step for the GGE is provided as follows:

Step 1: Normalization: Normalization of column C_i associated with the variable x_i is to force all nonzero coefficients of C_i has value of 1 for all positive numbers and -1 for all negative numbers and zero otherwise. It is equivalent to construct a diagonal matrix $D_i = \text{Diag}(d_{11}, d_{22}, \ldots, d_{(n-1)(n-1)}, d_m)$

Where $d_{kk} = 1$ if $l_{ki} = 0$ and $d_{kk} = 1/|l_{ki}|$ if $l_{ki} \neq 0$ such as $D_i f = D_i \vec{L}\vec{i} \geq 0$.

Note that Multiplying L by D_i is simply dividing the k-th row of L by 1 if $l_{ki} = 0$ and the j-th row of L by $|l_{ji}|$ if $l_{ji} \neq 0$. In other words, we have:

$$D_i L = \begin{bmatrix} l_{11}/d_{11} & l_{12}/d_{11} & \cdots & l_{1i}/d_{11} & \cdots & l_{1(m+1)}/d_{11} \\ l_{21}/d_{22} & l_{22}/d_{22} & \cdots & l_{2i}/d_{22} & \cdots & l_{2(m+1)}/d_{22} \\ l_{31}/d_{33} & l_{32}/d_{33} & \cdots & l_{3i}/d_{33} & \cdots & l_{3(m+1)}/d_{33} \\ \vdots & \vdots & \vdots & \vdots & \vdots & \vdots \\ l_{(n-1)1}/d_{(n-1)(n-1)} & l_{(n-1)1}/d_{(n-1)(n-1)} & \cdots & l_{(n-1)i}/d_{(n-1)(n-1)} & \cdots & l_{(n-1)(m+1)}/d_{(n-1)(n-1)} \\ l_{n1}/d_{nn} & l_{n2}/d_{nn} & \cdots & l_{ni}/d_{nn} & \cdots & l_{n(m+1)}/d_{nn} \end{bmatrix}$$

Note that we have the i-th column of $D_i L$ is

$$[l_{1i}/d_{11}, l_{2i}/d_{22}, \ldots, l_{ii}/d_{ii}, \ldots l_{ni}/d_{nn}]^T = [s_{1i}, s_{2i}, \ldots, s_{ii}, \ldots s_{ni}]^T = S_i$$

Where $s_{ki} = 1$ if $0 < l_{ki}$, $s_{ki} = -1$ if $l_{ki} < 0$, and $s_{ki} = 0$ if $l_{ki} = 0$, for $1 \leq k \leq n$.

Upon the completion of Step 1, the i-th column, associated with the variable, x_i consists of only three values, 1, -1, or 0. Step 2 is needed to group rows in $D_i L$ by the signs in S_i in its i-th column and also sort rows in each group by decreasing number of nonzero coefficients (nzc) of the associated row. In equation notations,

We are rearranging the rows of $D_i L$ into three parts by a row permutation matrix, P, such that

$$T = PD_i L = \begin{bmatrix} T_P \\ T_N \\ T_Z \end{bmatrix}$$ where $P = [e_{r_1}, e_{r_2}, ---, e_{r_n}]$ and the i-th column of

T_P is $\begin{bmatrix} 1 \\ 1 \\ \vdots \\ 1 \end{bmatrix}$ the i-th column of $T_N = \begin{bmatrix} -1 \\ -1 \\ \vdots \\ -1 \end{bmatrix}$, and the i-th column of $T_Z = \begin{bmatrix} 0 \\ 0 \\ \vdots \\ 0 \end{bmatrix}$.

Furthermore, we have

Step 3. Identifying most and least binding rows: Both T_P and T_N are sorted in decreasing order of $nzc(r)$ such that $r_k \leq r_j$ if $nzc(r_j) \geq nzc(r_k)$ where $nzc(r)$ where $nzc(r)$ is defined as the number of nonzero coefficients of the r-th row of T. We refer to rows in T_P or T_N with the maximum $nzc(r)$ as binding rows. Two cases of binding rows must be separated properly to locate the least and most binding rows. Let row r in T_P be, we have

$$\sum_{k=1}^{i-1}(-p_{rk})x_k + \sum_{k=i+1}^{m}(-p_{rk})x_k - p_{r(m+1)} = f_{-i}(\overrightarrow{x_{-i}}) \leq x_i$$

It is possible that another binding row s in T_P as:

$$\sum_{k=1}^{i-1}p_{sk}x_k + x_i + \sum_{k=i+1}^{m}p_{sk}x_k + p_{s(m+1)} \geq 0,$$
, we have

$$\sum_{k=1}^{i-1}(-p_{sk})x_k + \sum_{k=i+1}^{m}(-p_{sk})x_k - p_{s(m+1)} = g_{-i}(\overrightarrow{x_{-i}}) \leq x_i$$

Note that if $\sum_{k=1}^{i-1}(-p_{rk})x_k + \sum_{k=i+1}^{m}(-p_{rk})x_k = \sum_{k=1}^{i-1}(-p_{sk})x_k + \sum_{k=i+1}^{m}(-p_{sk})x_k = h_{-i}(\overrightarrow{x_{-i}})$, , we have

$$f_{-i}(\overrightarrow{x_{-i}}) = h_{-i}(\overrightarrow{x_{-i}}) - p_{r(m+1)} \leq x_i \quad \text{and} \quad g_{-i}(\overrightarrow{x_{-i}}) = h_{-i}(\overrightarrow{x_{-i}}) - p_{s(m+1)} \leq x_i,$$

, hence, the most lower binding (MLB) row for x_i is row j such that $p_{j(m+1)} = \max\{-p_{r(m+1)}, -p_{s(m+1)}, \dots, -p_{u(m+1)}\} = \min\{p_{r(m+1)}, p_{s(m+1)}, \dots, p_{u(m+1)}\}$

. Similarly, let row r in T_N be $\sum_{k=1}^{i-1}p_{rk}x_k - x_i + \sum_{k=i+1}^{m}p_{rk}x_k + p_{r(m+1)} \geq 0$, we have

$$x_i \leq \sum_{k=1}^{i-1}p_{rk}x_k + \sum_{k=i+1}^{m}p_{rk}x_k + p_{r(m+1)} = f_{-i}(\overrightarrow{x_{-i}})$$

It is possible that another binding row s in T_N as:

$$\sum_{k=1}^{i-1}p_{sk}x_k - x_i + \sum_{k=i+1}^{m}p_{sk}x_k + p_{s(m+1)} \geq 0,$$
,   we   have

$$x_i \leq \sum_{k=1}^{i-1}p_{sk}x_k + \sum_{k=i+1}^{m}p_{sk}x_k + p_{s(m+1)} = g_{-i}(\overrightarrow{x_{-i}})$$

Note that if $\sum_{k=1}^{i-1}p_{rk}x_k + \sum_{k=i+1}^{m}p_{rk}x_k = \sum_{k=1}^{i-1}p_{sk}x_k + \sum_{k=i+1}^{m}p_{sk}x_k = h_{-i}(\overrightarrow{x_{-i}})$, , we have

$$f_{-i}(\overrightarrow{x_{-i}}) = h_{-i}(\overrightarrow{x_{-i}}) + p_{r(m+1)} \leq x_i \quad \text{and} \quad g_{-i}(\overrightarrow{x_{-i}}) = h_{-i}(\overrightarrow{x_{-i}}) + p_{s(m+1)} \leq x_i,$$

hence, the least upper binding (LUB) row for x_i is row k such that $p_{k(m+1)} = \min\{p_{r(m+1)}, p_{s(m+1)}, \dots, p_{u(m+1)}\} = \max\{-p_{r(m+1)}, -p_{s(m+1)}, \dots, -p_{u(m+1)}\}$

Elimination of the variable x_i must preserve the most and/or least binding rows with respect to all the connected variables included and accounted for.

Step 4. Note that each row in T_P uniquely defines a lower bound of x_i with respect to all other variables in $\overline{x}_{-i} = (x_1, x_2, \ldots, x_{i-1}, x_{i+1}, \ldots x_m)$. Similarly, each row in T_N uniquely define an upper bound of x_i with respect to all other variables in x_{-i}.

Select distinct binding rows from T_P and T_N such that we have:

$$\sum_{k=1}^{i-1} p_{rk} x_k + x_i + \sum_{k=i+1}^{m} p_{rk} x_k + p_{r(m+1)} \geq 0 \text{ if row } r \text{ in } T_P$$

and

$$\sum_{k=1}^{i-1} p_{uk} x_k - x_i + \sum_{k=i+1}^{m} p_{uk} x_k + p_{u(m+1)} \geq 0 \text{ if row } u \text{ in } T_N$$

if row u in T_N

Consequently, we have a lower bound for x_i from a binding row in T_P as

$$\sum_{k=1}^{i-1} (-p_{rk}) x_k + \sum_{k=i+1}^{m} (-p_{rk}) x_k - p_{r(m+1)} = f_{-i}(\overrightarrow{x_{-i}}) \leq x_i$$

and an upper bound for x_i from a binding row in T_N as

$$x_i \leq g_{-i}(\overrightarrow{x_{-i}}) = \sum_{k=1}^{i-1} p_{uk} x_k + \sum_{k=i+1}^{m} p_{uk} x_k + p_{u(m+1)}$$

As shown in Step 3, a unique most lower binding (MLB) row and least upper binding (LUB) row for x_i may be identified if such binding relationship exists.

From this two extreme binding rows, the variable, x_i, can be safely and successfully eliminated as:

$$f_{-i}(\overrightarrow{x_{-i}}) \leq x_i \leq g_{-i}(\overrightarrow{x_{-i}}) \Rightarrow g_{-i}(\overrightarrow{x_{-i}}) - f_{-i}(\overrightarrow{x_{-i}}) = d_{-i}(\overrightarrow{x_{-i}}) \geq 0$$

For each row α within T_P we have

$$\sum_{k=1}^{i-1} p_{\alpha k} x_k + x_i + \sum_{k=i+1}^{m} p_{\alpha k} x_k + p_{\alpha(m+1)} \geq 0$$

$$x_i \geq v_{-i}(\overrightarrow{x_{-i}}) = \sum_{k=1}^{i-1} (-p_{\alpha k}) x_k + \sum_{k=i+1}^{m} (-p_{\alpha k}) x_k - p_{\alpha(m+1)}$$

i.e.,

Consequently, we have $f_{-i}(\overrightarrow{x_{-i}}) \geq x_i \geq v_{-i}(\overrightarrow{x_{-i}})$ i.e., $f_{-i}(\overrightarrow{x_{-i}}) - v_{-i}(\overrightarrow{x_{-i}}) = s_{-i}(\overrightarrow{x_{-i}}) \geq 0$

Similarly, for each row β within T_N we have

$$\sum_{k=1}^{i-1} p_{\beta k} x_k - x_i + \sum_{k=i+1}^{m} p_{\beta k} x_k + p_{\beta(m+1)} \geq 0$$

i.e.,

$$x_i \leq w_{-i}(x_{-i}) = \sum_{k=1}^{i-1}(p_{\beta k})x_k + \sum_{k=i+1}^{m}(p_{\beta k})x_k + p_{\beta(m+1)}$$

Consequently, we have $f_{-i}(x_{-i}) \leq x_i \leq w_{-i}(x_{-i})$ i.e., $w_{-i}(x_{-i}) - f_{-i}(x_{-i}) = t_{-x}(x_{-i}) \geq 0$

Note that rows in T_z do not contain the variable, x_i. In other words, we have shown that the variable, x_i may be eliminated completely and safely from every row in T without loss of all the binding inequalities that include all possible lower or upper bounds in terms of the remaining variables, $\overline{x_{-i}}$

Note that steps 1 to 4 may be applied recursively to eliminate any specific ordering of variables in $\vec{x}$ such that only a specific variable, x_j left such that we have the final inequalities:

$$
\begin{bmatrix} x_1 & . & x_{j-1} & x_j & x_{j+1} & . & x_{n-1} & x_n & 1 \end{bmatrix} \quad \vec{t}
$$

$$
\begin{bmatrix}
0 & \cdots & 0 & 1 & 0 & \cdots & 0 & 0 & p_1 \\
: & \cdots & 0 & : & 0 & \cdots & 0 & 0 & : \\
0 & \cdots & 0 & 1 & 0 & \cdots & 0 & 0 & p_{|T_p|} \\
0 & \cdots & 0 & -1 & 0 & \cdots & 0 & 0 & q_1 \\
0 & \cdots & 0 & : & 0 & \cdots & 0 & 0 & : \\
0 & \cdots & 0 & -1 & 0 & \cdots & 0 & 0 & q_{|T_N|} \\
0 & \cdots & 0 & 0 & 0 & \cdots & 0 & 0 & r_1 \\
0 & \cdots & 0 & : & 0 & \cdots & 0 & 0 & : \\
0 & \cdots & 0 & 0 & 0 & \cdots & 0 & 0 & r_{|V_z|}
\end{bmatrix}
\begin{bmatrix}
x_1 \\ : \\ x_{j-1} \\ x_j \\ x_{j+1} \\ : \\ x_{n-1} \\ x_n \\ 1
\end{bmatrix} \geq 0
$$

(5)

Inequalities (5) define all possible upper bounds and lower bounds for the select variables, x_j.

In other words, we have:

$$-p_k \leq x_j \leq q_m \ \forall k = 1, 2, \ldots, |T_p|, m = 1, 2, \ldots, |T_N|$$ where $|T_p|$ is the number of rows in T_p and T_N is the number of rows in T_N

Let $$p^* = \max_{1 \leq k \leq |T_p|}\{-p_k\} \leq x_j \leq q^* = \min_{1 \leq m \leq |T_N|}\{q_m\}$$, then $F_j = [p^*, q^*]$ uniquely defines a feasible interval for the variable, x_j. Consequently, we may have the following 6 possible cases for $F_j = [p^*, q^*]$.

(i) no solution such that $F_j = [p^*, q^*]$ if $q^* < p^*$

(ii) unique solution with $F_j = [p^*, q^*]$ with $p^* = q^* = r$ i.e., we have $x_j = r$.

(iii) bounded interval with infinite many solutions as $F_j = [p^*, q^*]$ if $p^* < q^*$ & $-\infty < p^* < q^* < \infty$.

(iv) bounded above only as $F_j = (-\infty, q^*]$ if $|T_P| = \phi$.

(v) bounded below only as $F_j = [p^*, \infty)$ if $|T_N| = \phi$.

(vi) unbounded above and below as $F_j = (-\infty, \infty)$ if $|T_P| = \phi$ and $|T_N| = \phi$.

EXAMPLES

To illustrate the validity, capability, and correctness of GGE, we provide feasible intervals for simple inequalities computed by GGE that are easily verifiable and sample linear programs covering with unique solution, no solution, or unbounded cases as follows:

Example #1

L1:	$x - 3y \le 6$
L2:	$3x + 4y \le 12$
L3:	$2x + y \ge 4$

$$(6)$$

The homogeneous $L*t \ge 0$ is:

$$\begin{bmatrix} -1 & 3 & 6 \\ -3 & -4 & 12 \\ 2 & 1 & -4 \end{bmatrix} * \begin{bmatrix} x \\ y \\ 1 \end{bmatrix} \ge \begin{bmatrix} 0 \\ 0 \\ 0 \end{bmatrix}$$

$$(7)$$

Eliminating x, we have the following inequalities

$$\begin{bmatrix} 0 & -5/2 & 6 \\ 0 & -11/2 & 18 \\ 0 & 7 & 8 \end{bmatrix} * \begin{bmatrix} x \\ y \\ 1 \end{bmatrix} \ge \begin{bmatrix} 0 \\ 0 \\ 0 \end{bmatrix}$$

$$(8)$$

This provides the feasible interval for x as $F_y = [-8/7, 12/5]$

Eliminating y, we have the following inequalities

$$\begin{bmatrix} -13/4 & 0 & 60/4 \\ -13/3 & 0 & 20 \\ 5/4 & 0 & -1 \end{bmatrix} * \begin{bmatrix} x \\ y \\ 1 \end{bmatrix} \geq \begin{bmatrix} 0 \\ 0 \\ 0 \end{bmatrix}$$

(9)

This provides the feasible interval for y as $F_x = [4/5, 60/13]$

Example #2

L1: $2y + 3z \leq 5$

L2: $x + y + 2z \leq 4$

L3: $x + 2y + 3z \leq 7$ (10)

L4: $x \geq 0$

L5: $y \geq 0$

L6: $z \geq 0$

The homogeneous $L * t \geq 0$ is:

$$\begin{bmatrix} 0 & -2 & -3 & 5 \\ -1 & -1 & -2 & 4 \\ -1 & -2 & -3 & 7 \\ 1 & 0 & 0 & 0 \\ 0 & 1 & 0 & 0 \\ 0 & 0 & 1 & 0 \end{bmatrix} * \begin{bmatrix} x \\ y \\ z \\ 1 \end{bmatrix} \geq \begin{bmatrix} 0 \\ 0 \\ 0 \\ 0 \\ 0 \\ 0 \end{bmatrix}$$

(11)

Eliminating variables y and z with (11), we have the following inequalities:

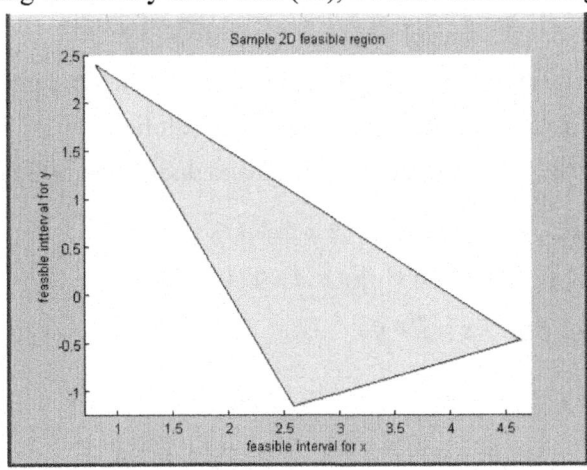

Figure 1: Feasible Intervals F_x and F_y

$$
\begin{bmatrix}
0 & 0 & -3 & 5 \\
-1 & 0 & -2 & 4 \\
-1 & 0 & -3 & 7 \\
1 & 0 & 0 & 0 \\
-1 & 0 & -2 & 4 \\
0 & 0 & 1 & 0
\end{bmatrix}
*
\begin{bmatrix} x \\ y \\ z \\ 1 \end{bmatrix}
\geq
\begin{bmatrix} 0 \\ 0 \\ 0 \\ 0 \\ 0 \\ 0 \end{bmatrix}
\text{ and }
\begin{bmatrix}
0 & 0 & 0 & 5 \\
-1 & 0 & 0 & 4 \\
-1 & 0 & 0 & 7 \\
1 & 0 & 0 & 0 \\
-1 & 0 & 0 & 0 \\
-1/2 & 0 & 0 & 0
\end{bmatrix}
*
\begin{bmatrix} x \\ y \\ z \\ 1 \end{bmatrix}
\geq
\begin{bmatrix} 0 \\ 0 \\ 0 \\ 0 \\ 0 \\ 0 \end{bmatrix}
\tag{12}
$$

This provides the feasible interval for x as $F_x = [0,4]$

Eliminating variable x and y with (11), we have the following inequalities:

$$
\begin{bmatrix}
0 & -2 & -3 & 5 \\
0 & -1 & -2 & 4 \\
0 & -2 & -3 & 7 \\
0 & -1 & -2 & 4 \\
0 & 1 & 0 & 0 \\
0 & 0 & 1 & 0
\end{bmatrix}
*
\begin{bmatrix} x \\ y \\ z \\ 1 \end{bmatrix}
\geq
\begin{bmatrix} 0 \\ 0 \\ 0 \\ 0 \\ 0 \\ 0 \end{bmatrix}
\text{ and }
\begin{bmatrix}
0 & 0 & -3 & 5 \\
0 & 0 & -2 & 4 \\
0 & 0 & -3 & 7 \\
0 & 0 & -2 & 4 \\
0 & 0 & -3/2 & 7/2 \\
0 & 0 & 1 & 0
\end{bmatrix}
*
\begin{bmatrix} x \\ y \\ z \\ 1 \end{bmatrix}
\geq
\begin{bmatrix} 0 \\ 0 \\ 0 \\ 0 \\ 0 \\ 0 \end{bmatrix}
\tag{13}
$$

This provides the feasible interval for z as $F_z = [0,5/3]$

Eliminating variables x and z with (11), we have the following inequalities:

$$
\begin{bmatrix}
0 & -2 & -3 & 5 \\
0 & -1 & -2 & 4 \\
0 & -2 & -3 & 7 \\
0 & -1 & -2 & 4 \\
0 & 1 & 0 & 0 \\
0 & 0 & 1 & 0
\end{bmatrix}
*
\begin{bmatrix} x \\ y \\ z \\ 1 \end{bmatrix}
\geq
\begin{bmatrix} 0 \\ 0 \\ 0 \\ 0 \\ 0 \\ 0 \end{bmatrix}
\text{ and }
\begin{bmatrix}
0 & -2 & 0 & 5 \\
0 & -1 & 0 & 4 \\
0 & -2 & 0 & 7 \\
0 & -1 & 0 & 4 \\
0 & 1 & 0 & 0 \\
0 & -1 & 0 & 4
\end{bmatrix}
*
\begin{bmatrix} x \\ y \\ z \\ 1 \end{bmatrix}
\geq
\begin{bmatrix} 0 \\ 0 \\ 0 \\ 0 \\ 0 \\ 0 \end{bmatrix}
\tag{14}
$$

This provides the feasible interval for y as $F_y = [0,5/2]$

Figure 6 illustrates the feasible region for these feasible intervals.

Example #3 a linear program with unique solution:

Consider the following pair of primal and dual linear programs:

Primary LP: $Max \ \{c \bullet x \,|\, A*x \leq b; l \leq x \leq u\}$

Dual LP: $Min \ \{b*y \,|\, y*A = c; 0 \leq y\}$

Optimality: $c*x = b*y$

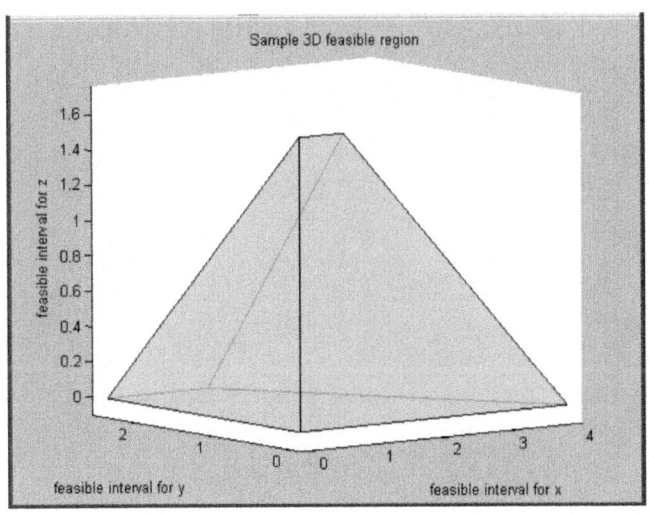

Figure 6: Feasible Intervals for F_x, F_y, and F_z

The self-dual LP as Homogeneous Liner Inequalities (HLI) (Wang, 2013):

$$\begin{bmatrix} 0 & -A & b \\ A^T & 0 & -c \\ -b & c & 0 \\ b & -c & 0 \\ I & 0 & 0 \\ 0 & I & -l \\ 0 & -I & u \end{bmatrix} * \begin{bmatrix} y \\ x \\ 1 \end{bmatrix} = S*t = f \geq 0$$

Consider the primal and dual LP pair:

$$\max z = -12x_1 - 18x_2 - 15x_3$$

$$\text{subject to:} \quad \begin{bmatrix} -4 & -8 & -6 \\ -3 & -6 & -12 \end{bmatrix} * \begin{bmatrix} x_1 \\ x_2 \\ x_3 \end{bmatrix} \leq \begin{bmatrix} -64 \\ -96 \end{bmatrix}, \quad 0 \leq \begin{bmatrix} x_1 \\ x_2 \\ x_3 \end{bmatrix}$$

$$\min w = -64y_1 - 96y_2$$

$$\text{subject to:} \quad \begin{bmatrix} -4 & -3 \\ -8 & -6 \\ -6 & -12 \end{bmatrix} * \begin{bmatrix} y_1 \\ y_2 \end{bmatrix} \geq \begin{bmatrix} -12 \\ -18 \\ -15 \end{bmatrix}, \quad 0 \leq \begin{bmatrix} y_1 \\ y_2 \end{bmatrix}$$

The corresponding HLFS for this LP as self-dual form [18, 19] is:

$$
\begin{bmatrix} y_1 & y_2 & x_1 & x_2 & x_3 & 1 \end{bmatrix}
$$

	y_1	y_2	x_1	x_2	x_3	1
r_1	0	0	2	4	3	−32
r_2	0	0	1	2	4	−32
r_3	−4	−3	0	0	0	12
.	−8	−6	0	0	0	18
r_5	−6	−12	0	0	0	15
r_6	64	96	−12	−18	−15	0
.	−64	−96	12	18	15	0
.	1	0	0	0	0	0
.	0	1	0	0	0	0
.	0	0	1	0	0	0
r_{11}	0	0	0	1	0	0
r_{12}	0	0	0	0	1	0

$$
*\begin{bmatrix} y_2 \\ y_1 \\ x_1 \\ x_2 \\ x_3 \\ 1 \end{bmatrix} \geq \begin{bmatrix} 0 \\ 0 \\ 0 \\ 0 \\ 0 \\ 0 \\ 0 \\ 0 \\ 0 \\ 0 \\ 0 \\ 0 \end{bmatrix}
$$

(15)

Applying GGE to eliminate y_1, normalize column y_1 we have:

$$
\begin{bmatrix} y_1 & y_2 & x_1 & x_2 & x_3 & 1 \end{bmatrix}
$$

	y_1	y_2	x_1	x_2	x_3	1
r_1	0	0	2	4	3	−32
r_2	0	0	1	2	4	−32
r_3	−1	−3/4	0	0	0	13
.	−1	−3/4	0	0	0	9/4
r_5	−1	−2	0	0	0	5/2
r_6	1	3/2	−3/16	−9/32	−15/64	0
.	−1	−3/2	3/16	9/32	15/64	0
.	1	0	0	0	0	0
.	0	1	0	0	0	0
.	0	0	1	0	0	0
r_{11}	0	0	0	1	0	0
r_{12}	0	0	0	0	1	0

$$
*\begin{bmatrix} y_2 \\ y_1 \\ x_1 \\ x_2 \\ x_3 \\ 1 \end{bmatrix} \geq \begin{bmatrix} 0 \\ 0 \\ 0 \\ 0 \\ 0 \\ 0 \\ 0 \\ 0 \\ 0 \\ 0 \\ 0 \\ 0 \end{bmatrix}
$$

(16)

Select distinct binding rows, LUB r_5 and MLB r_6 to eliminate y_1, we Have

$$
\begin{array}{c}
r \\ r \\ r \\ : \\ r_5 \\ r_6 \\ . \\ .. \\ . \\ . \\ r_{11} \\ r_{12}
\end{array}
\begin{bmatrix}
0 & 0 & 2 & 4 & 3 & -32 \\
0 & 0 & 1 & 2 & 4 & -32 \\
0 & 3/4 & -3/16 & -9/32 & -15/64 & 3 \\
0 & 3/4 & -3/16 & -9/32 & -15/64 & 9/4 \\
0 & -1/2 & -3/16 & -9/32 & -15/84 & 5/2 \\
0 & -1/2 & -3/16 & -9/32 & -15/64 & 5/2 \\
0 & 0 & 0 & 0 & 0 & 0 \\
0 & -2 & 0 & 0 & 0 & 5/2 \\
0 & 1 & 0 & 0 & 0 & 0 \\
0 & 0 & 1 & 0 & 0 & 0 \\
0 & 0 & 0 & 1 & 0 & 0 \\
0 & 0 & 0 & 0 & 1 & 0
\end{bmatrix}
*
\begin{bmatrix}
y_2 \\ y_1 \\ x_1 \\ x_2 \\ x_3 \\ 1
\end{bmatrix}
\geq
\begin{bmatrix}
0 \\ 0 \\ 0 \\ 0 \\ 0 \\ 0 \\ 0 \\ 0 \\ 0 \\ 0 \\ 0 \\ 0
\end{bmatrix}
\qquad (17)
$$

(columns: y_1, y_2, x_1, x_2, x_3, 1)

After normalization for of 2th column for y_2, we have:

$$
\begin{array}{c}
r \\ r \\ r \\ : \\ r_5 \\ r_6 \\ . \\ .. \\ . \\ . \\ r_{11} \\ r_{12}
\end{array}
\begin{bmatrix}
0 & 0 & 2 & 4 & 3 & -32 \\
0 & 0 & 1 & 2 & 4 & -32 \\
0 & 1 & -1/4 & -3/8 & -5/16 & 4 \\
0 & 1 & -1/4 & -3/8 & -5/16 & 3 \\
0 & -1 & -3/8 & -9/16 & -15/32 & 5 \\
0 & -1 & -3/8 & -9/16 & -15/32 & 5 \\
0 & 0 & 0 & 0 & 0 & 0 \\
0 & -1 & 0 & 0 & 0 & 5/4 \\
0 & 1 & 0 & 0 & 0 & 0 \\
0 & 0 & 1 & 0 & 0 & 0 \\
0 & 0 & 0 & 1 & 0 & 0 \\
0 & 0 & 0 & 0 & 1 & 0
\end{bmatrix}
*
\begin{bmatrix}
y_2 \\ y_1 \\ x_1 \\ x_2 \\ x_3 \\ 1
\end{bmatrix}
\geq
\begin{bmatrix}
0 \\ 0 \\ 0 \\ 0 \\ 0 \\ 0 \\ 0 \\ 0 \\ 0 \\ 0 \\ 0 \\ 0
\end{bmatrix}
\qquad (18)
$$

(columns: y_1, y_2, x_1, x_2, x_3, 1)

Select distinct binding rows, MLB r_4 and LUB r_5 to eliminate y_2, we have

$$
\begin{array}{c}
r_1 \\ r_2 \\ r_3 \\ r_4 \\ r_5 \\ r_6 \\ r_7 \\ r_8 \\ r_9 \\ r_{10} \\ r_{11} \\ r_{12}
\end{array}
\begin{matrix}
y_1 & y_2 & x_1 & x_2 & x_3 & 1 \\
\left[\begin{array}{cccccc}
0 & 0 & 2 & 4 & 3 & -32 \\
0 & 0 & 1 & 2 & 4 & -32 \\
0 & 0 & -5/8 & -15/16 & -25/32 & 9 \\
0 & 0 & -5/8 & -15/16 & -25/32 & 8 \\
0 & 0 & -5/8 & -15/16 & -25/32 & 8 \\
0 & 0 & -5/8 & -15/16 & -25/32 & 8 \\
0 & 0 & 0 & 0 & 0 & 0 \\
0 & 0 & -1/4 & -3/8 & -5/16 & 17/4 \\
0 & 0 & -3/8 & -9/16 & -15/32 & 5 \\
0 & 0 & 1 & 0 & 0 & 0 \\
0 & 0 & 0 & 1 & 0 & 0 \\
0 & 0 & 0 & 0 & 1 & 0
\end{array}\right]
\end{matrix}
*
\begin{bmatrix}
y_2 \\ y_1 \\ x_1 \\ x_2 \\ x_3 \\ 1
\end{bmatrix}
\geq
\begin{bmatrix}
0 \\ 0 \\ 0 \\ 0 \\ 0 \\ 0 \\ 0 \\ 0 \\ 0 \\ 0 \\ 0 \\ 0
\end{bmatrix}
\tag{19}
$$

After normalization for of 3rd column and the removal of identical rows, we have:

$$
\begin{array}{c}
r_1 \\ r_2 \\ r_3 \\ . \\ . \\ r_6 \\ r_7 \\ r_8 \\ r_9
\end{array}
\begin{matrix}
y_1 & y_2 & x_1 & x_2 & x_3 & 1 \\
\left[\begin{array}{cccccc}
0 & 0 & 1 & 2 & 3/2 & -16 \\
0 & 0 & 1 & 2 & 4 & -32 \\
0 & 0 & -1 & -3/2 & -5/4 & 72/5 \\
0 & 0 & -1 & -3/2 & -5/4 & 64/5 \\
0 & 0 & -1 & -3/2 & -5/4 & 17 \\
0 & 0 & -1 & -3/2 & -5/4 & 40/3 \\
0 & 0 & 1 & 0 & 0 & 0 \\
0 & 0 & 0 & 1 & 0 & 0 \\
0 & 0 & 0 & 0 & 1 & 0
\end{array}\right]
\end{matrix}
*
\begin{bmatrix}
y_2 \\ y_1 \\ x_1 \\ x_2 \\ x_3 \\ 1
\end{bmatrix}
\geq
\begin{bmatrix}
0 \\ 0 \\ 0 \\ 0 \\ 0 \\ 0 \\ 0 \\ 0 \\ 0
\end{bmatrix}
\tag{20}
$$

Select distinct binding rows, MLB r_1 and LUB r_4 to eliminate x_1, we have

$$
\begin{array}{c}
\\
r_1 \\
r_2 \\
r_3 \\
. \\
. \\
r_6 \\
r_7 \\
r_8 \\
r_9
\end{array}
\begin{array}{cccccc}
y_1 & y_2 & x_1 & x_2 & x_3 & 1 \\
\left[\begin{array}{cccccc}
0 & 0 & 0 & 1/2 & 1/4 & -16/5 \\
0 & 0 & 0 & 1/2 & 11/4 & -96/5 \\
0 & 0 & 0 & 1/2 & 1/4 & -8/5 \\
0 & 0 & 0 & 1/2 & 1/4 & -16/5 \\
0 & 0 & 0 & 1/2 & 1/4 & 1 \\
0 & 0 & 0 & 1/2 & 1/4 & -8/3 \\
0 & 0 & 0 & -3/2 & -5/4 & 64/5 \\
0 & 0 & 0 & 1 & 0 & 0 \\
0 & 0 & 0 & 0 & 1 & 0
\end{array}\right]
\end{array}
*
\begin{bmatrix}
y_2 \\
y_1 \\
x_1 \\
x_2 \\
x_3 \\
1
\end{bmatrix}
\geq
\begin{bmatrix}
0 \\
0 \\
0 \\
0 \\
0 \\
0 \\
0 \\
0 \\
0
\end{bmatrix}
\tag{21}
$$

After normalization for of 4th column for x_2 and the removal of identical rows, we have:

$$
\begin{array}{c}
\\
r_1 \\
r_2 \\
r_3 \\
. \\
. \\
r_6 \\
r_7 \\
r_8 \\
r_9
\end{array}
\begin{array}{cccccc}
y_1 & y_2 & x_1 & x_2 & x_3 & 1 \\
\left[\begin{array}{cccccc}
0 & 0 & 0 & 1 & 1/2 & -32/5 \\
0 & 0 & 0 & 1 & 11/2 & -192/5 \\
0 & 0 & 0 & 1 & 1/2 & -16/5 \\
0 & 0 & 0 & 1 & 1/2 & -32/5 \\
0 & 0 & 0 & 1 & 1/2 & 2 \\
0 & 0 & 0 & 1 & 1/2 & -16/3 \\
0 & 0 & 0 & -1 & -5/6 & 128/15 \\
0 & 0 & 0 & 1 & 0 & 0 \\
0 & 0 & 0 & 0 & 1 & 0
\end{array}\right]
\end{array}
*
\begin{bmatrix}
y_2 \\
y_1 \\
x_1 \\
x_2 \\
x_3 \\
1
\end{bmatrix}
\geq
\begin{bmatrix}
0 \\
0 \\
0 \\
0 \\
0 \\
0 \\
0 \\
0 \\
0
\end{bmatrix}
\tag{22}
$$

Select distinct binding rows, MLB r_1 and LUB r_7 to eliminate x_2, we have

$$
\begin{array}{c}
\\
r_1 \\
r_2 \\
r_3 \\
r_4 \\
r_5 \\
r_6 \\
r_7
\end{array}
\begin{array}{cccccc}
y_1 & y_2 & x_1 & x_2 & x_3 & 1 \\
\left[\begin{array}{cccccc}
0 & 0 & 0 & 0 & -2/6 & 32/15 \\
0 & 0 & 0 & 0 & 7/3 & -224/15 \\
0 & 0 & 0 & 0 & -2/6 & 16/3 \\
0 & 0 & 0 & 0 & -2/6 & 98/15 \\
0 & 0 & 0 & 0 & -2/6 & 32/15 \\
0 & 0 & 0 & 0 & -5/6 & 128/15 \\
0 & 0 & 0 & 0 & 1 & 0
\end{array}\right]
\end{array}
*
\begin{bmatrix}
y_2 \\
y_1 \\
x_1 \\
x_2 \\
x_3 \\
1
\end{bmatrix}
\geq
\begin{bmatrix}
0 \\
0 \\
0 \\
0 \\
0 \\
0 \\
0
\end{bmatrix}
\tag{23}
$$

After normalization for of 5th column for x_3 and the removal of identical rows, we have:

$$
\begin{array}{c}
 \\
r_1 \\
r_2 \\
r_3 \\
r_4 \\
r_5 \\
r_6
\end{array}
\begin{array}{cccccc}
y_1 & y_2 & x_1 & x_2 & x_3 & 1 \\
\end{array}
\begin{bmatrix}
0 & 0 & 0 & 0 & -1 & 32/5 \\
0 & 0 & 0 & 0 & 1 & -32/5 \\
0 & 0 & 0 & 0 & -1 & 16 \\
0 & 0 & 0 & 0 & -1 & 98/15 \\
0 & 0 & 0 & 0 & -1 & 128/15 \\
0 & 0 & 0 & 0 & 1 & 0
\end{bmatrix}
*
\begin{bmatrix}
y_2 \\
y_1 \\
x_1 \\
x_2 \\
x_3 \\
1
\end{bmatrix}
\geq
\begin{bmatrix}
0 \\
0 \\
0 \\
0 \\
0 \\
0
\end{bmatrix}
$$

(24)

Hence the MLB is obtained from row r_1 and the LUB is obtained from row r_2 such that

$$\text{MLB} = \max\{0, 32/5\} \text{ and } \text{LUB} = \min\{32/5, 98/5, 128/5, 16\} = 6.4$$

Since MLB=LUB=6.4, we conclude that the variable x_3 has unique solution with $x_3 = 6.4$

From (21) with $x_3 = 6.4$, for x_2 the MLB is $\max\{-5.2, 2.13333, 3.2\} = 3.2$ and LUB =3.2

Hence, x_2 has unique solution $x_2 = 3.2$

Substituting $x_3 = 6.4$ and $x_2 = 3.2$ into (19), we obtain the MLB and LUB for x_1 as MLB = 0 and

$$\text{LUB} = \min\{0, 0.53333, 1.5, 4.2\} = 0; \text{ Hence, } x_1 \text{ has unique solution } x_1 = 0.$$

Substituting $x_3 = 6.4, x_2 = 3.2$ and $x_1 = 0$ into (17), we obtain the MLB and LUB for y_2 as

$$\text{MLB} = \max\{0.8, 0, 0.2\} = 0.2 \text{ and LUB} = \min\{0.2, 1.25\} = 0.2; \text{ Hence, } y_2 \text{ has the}$$
unique solution = 0.2.

Substituting, $x_3 = 6.4 \; x_2 = 3.2, \; x_1 = 0$, and $y_2 = 0.2$ into (15), we obtain the MLB and LUB for y_1 as

$$\text{MLB} = \max\{0, 2.1\} = 2.1 \text{ and LUB} = \min\{2.1, 12.85\} = 2.1; \text{ Hence, } y_1 \text{ has}$$
unique solution $y_1 = 2.1$.

Consequently, both the primal and the dual LPs are resolved by applying the GGE algorithm to (15) with the unique optimal solution of $y_1 = 2.1, \; y_2 = 0.2, \; x_1 = 0, \; x_2 = 3.2, \; \& \; x_3 = 6.4$

Example #4 LP with no solution

Consider the LP:

$$\max \ z = x_1 + x_2 \quad \text{subject to:}$$

$$\begin{bmatrix} -1 & 2 \\ 2 & 1 \end{bmatrix} * \begin{bmatrix} x_1 \\ x_2 \end{bmatrix} \leq \begin{bmatrix} -6 \\ 4 \end{bmatrix} \qquad x_1 \geq 0 \ \text{ and } \ x_2 \geq 0$$

The corresponding HLFS for this LP as self-dual form [19] is:

	y_1	y_2	x_1	x_2	1				
r_1	0	0	1	-2	-6	*	y_1	≥	0
r_2	0	0	-2	-1	4		y_2		0
r_3	-1	2	0	0	-1		x_1		0
r_4	2	1	0	0	-1		x_2		0
r_5	6	-4	1	1	0		1		0
r_6	-6	4	-1	-1	0				0
r_7	1	0	0	0	0				0
r_8	0	1	0	0	0				0
r_9	0	0	1	0	0				0
r_{10}	0	0	0	1	0				0

(25)

Applying the GGE algorithm to normalize the first column, we have

	y_1	y_2	x_1	x_2	1				
r_1	0	0	1	-2	-6	*	y_1	≥	0
r_2	0	0	-2	-1	4		y_2		0
r_3	-1	2	0	0	-1		x_1		0
r_4	1	1/2	0	0	-1/2		x_2		0
r_5	1	-2/3	1/6	1/6	0		1		0
r_6	-1	2/3	-1/6	-1/6	0				0
r_7	1	0	0	0	0				0
r_8	0	1	0	0	0				0
r_9	0	0	1	0	0				0
r_{10}	0	0	0	1	0				0

(26)

Eliminate variable y_1 by rows r_6 and r_4, we have

$$
\begin{array}{c}
\begin{array}{ccccc} y_1 & y_2 & x_1 & x_2 & 1 \end{array} \\
\begin{array}{c} r_1 \\ r_2 \\ r_3 \\ r_4 \\ r_5 \\ r_6 \\ r_7 \end{array}
\begin{bmatrix}
0 & 0 & 1 & -2 & -6 \\
0 & 0 & -2 & -1 & 4 \\
0 & 5 & 0 & 0 & -3 \\
0 & 7 & -1 & -1 & -3 \\
0 & 1 & -1/4 & -1/4 & 0 \\
0 & 0 & 1 & 0 & 0 \\
0 & 0 & 0 & 1 & 0
\end{bmatrix}^*
\begin{bmatrix} y_1 \\ y_2 \\ x_1 \\ x_2 \\ 1 \end{bmatrix}
\overset{\geq}{}
\begin{bmatrix} 0 \\ 0 \\ 0 \\ 0 \\ 0 \\ 0 \\ 0 \end{bmatrix}
\end{array}
\tag{27}
$$

The 2nd column for variable y_2 are all positive, hence y_2 is only bounded below, the dual LP is unbounded above. For the primal LP, it is reduced to the following linear inequalities:

$$
\begin{array}{c}
\begin{array}{ccc} x_1 & x_2 & 1 \end{array} \\
\begin{array}{c} r_1 \\ r_2 \\ r_3 \\ r_4 \end{array}
\begin{bmatrix}
1 & -2 & -6 \\
-2 & -1 & 4 \\
1 & 0 & 0 \\
0 & 1 & 0
\end{bmatrix}^*
\begin{bmatrix} x_1 \\ x_2 \\ 1 \end{bmatrix}
\overset{\geq}{}
\begin{bmatrix} 0 \\ 0 \\ 0 \\ 0 \end{bmatrix}
\end{array}
\tag{28}
$$

Using GGE from rows r_1 and r_2, we can eliminate x_1 to obtain the both MLB and LUB for x_2 as $0 \le x_2$ and $x_2 \le -8/5$; Consequently, the feasible interval for x_2 is $f_{x_2} = \phi$, i.e., the primal LP has no solution !

Example #5 LP with unbounded solution

Consider the LP:

$$
\max z = 2x_1 - x_2 \quad \text{subject to:}
$$

$$
\begin{bmatrix} 1 & -1 \\ -2 & -1 \end{bmatrix} * \begin{bmatrix} x_1 \\ x_2 \end{bmatrix} \le \begin{bmatrix} 1 \\ -6 \end{bmatrix} \qquad x_1 \ge 0 \text{ and } x_2 \ge 0
$$

The corresponding HLFS for this LP as self-dual form [18] is:

$$
\begin{array}{c}
 \\
r_1 \\
r_2 \\
r_3 \\
r_4 \\
r_5 \\
r_6 \\
r_7 \\
r_8 \\
r_9 \\
r_{10}
\end{array}
\begin{bmatrix}
y_1 & y_2 & x_1 & x_2 & 1 \\
0 & 0 & -1 & 1 & 1 \\
0 & 0 & 2 & 1 & -6 \\
1 & -2 & 0 & 0 & -2 \\
-1 & -1 & 0 & 0 & 1 \\
-1 & 6 & 2 & -1 & 0 \\
1 & -6 & -2 & 1 & 0 \\
1 & 0 & 0 & 0 & 0 \\
0 & 1 & 0 & 0 & 0 \\
0 & 0 & 1 & 0 & 0 \\
0 & 0 & 0 & 1 & 0
\end{bmatrix}
*
\begin{bmatrix}
y_1 \\
y_2 \\
x_1 \\
x_2 \\
1
\end{bmatrix}
\geq
\begin{bmatrix}
0 \\
0 \\
0 \\
0 \\
0 \\
0 \\
0 \\
0 \\
0 \\
0
\end{bmatrix}
\tag{29}
$$

Using GGE from r_3 and r_5, we can eliminate y_1 and obtain

$$
\begin{array}{c}
 \\
r_1 \\
r_2 \\
r_3 \\
r_4 \\
r_5 \\
r_6 \\
r_7 \\
r_8
\end{array}
\begin{bmatrix}
y_1 & y_2 & x_1 & x_2 & 1 \\
0 & 0 & -1 & 1 & 1 \\
0 & 0 & 2 & 1 & -6 \\
0 & 4 & 2 & -1 & -2 \\
0 & -3 & 0 & 0 & -1 \\
0 & 6 & 2 & -1 & 0 \\
0 & 1 & 0 & 0 & 0 \\
0 & 0 & 1 & 0 & 0 \\
0 & 0 & 0 & 1 & 0
\end{bmatrix}
*
\begin{bmatrix}
y_1 \\
y_2 \\
x_1 \\
x_2 \\
1
\end{bmatrix}
\geq
\begin{bmatrix}
0 \\
0 \\
0 \\
0 \\
0 \\
0 \\
0 \\
0
\end{bmatrix}
\tag{30}
$$

Normalize the 2nd column for y_1, we have

$$
\begin{array}{c}
 \\
r_1 \\
r_2 \\
r_3 \\
r_4 \\
r_5 \\
r_6 \\
r_7 \\
r_8
\end{array}
\begin{array}{cccccc}
y_1 & y_2 & x_1 & x_2 & 1 \\
\left[\begin{array}{ccccc}
0 & 0 & -1 & 1 & 1 \\
0 & 0 & 2 & 1 & -6 \\
0 & 1 & 1/2 & -1/4 & -1/2 \\
0 & -1 & 0 & 0 & -1/3 \\
0 & 1 & 1/3 & -1/6 & 0 \\
0 & 1 & 0 & 0 & 0 \\
0 & 0 & 1 & 0 & 0 \\
0 & 0 & 0 & 1 & 0
\end{array}\right]
\end{array}
{}^{*}
\left[\begin{array}{c}
y_1 \\
y_2 \\
x_1 \\
x_2 \\
1
\end{array}\right]
\overset{\geq}{}
\left[\begin{array}{c}
0 \\
0 \\
0 \\
0 \\
0 \\
0 \\
0 \\
0
\end{array}\right]
$$

$$\tag{31}$$

Regardless the primal variables x_1 and x_2, the 2nd column for y_2 and the last column for MLB and LUB,

We obtain the feasible interval $f_{y_2} = \phi$ (i.e., $f = [\alpha, \beta] = \phi$ if $\beta < \alpha$) from rows r_4 and r_6 as $y_2 \leq -1/3 = \beta$ and $\alpha = 0 \leq y_2$

Eliminate y_2 by rows r_3 and r_4 and excluding the rows with dual variables y_1 and y_2 only, we have inequalities for the primal variables x_1 and x_2 as:

$$
\begin{array}{c}
 \\
r_1 \\
r_2 \\
r_3 \\
r_4 \\
r_5 \\
r_6
\end{array}
\begin{array}{cccc}
x_1 & x_2 & 1 \\
\left[\begin{array}{ccc}
-1 & 1 & 1 \\
2 & 1 & -6 \\
1/2 & -1/4 & -5/6 \\
1/3 & -1/6 & -1/3 \\
1 & 0 & 0 \\
0 & 1 & 0
\end{array}\right]
\end{array}
{}^{*}
\left[\begin{array}{c}
x_1 \\
x_2 \\
1
\end{array}\right]
\overset{\geq}{}
\left[\begin{array}{c}
0 \\
0 \\
0 \\
0 \\
0 \\
0
\end{array}\right]
$$

$$\tag{32}$$

Normalize the 3rd column for x_1, we have:

$$
\begin{array}{c}
 \\
r_1 \\
r_2 \\
r_3 \\
r_4 \\
r_5 \\
r_6
\end{array}
\begin{array}{c}
\begin{array}{ccc}
x_1 & x_2 & 1
\end{array} \\
\left[
\begin{array}{ccc}
-1 & 1 & 1 \\
1 & 1/2 & -3 \\
1 & -1/2 & -5/3 \\
1 & -1/2 & -1 \\
1 & 0 & 0 \\
0 & 1 & 0
\end{array}
\right]
\end{array}
*
\begin{bmatrix}
x_1 \\
x_2 \\
1
\end{bmatrix}
\geq
\begin{bmatrix}
0 \\
0 \\
0 \\
0 \\
0 \\
0
\end{bmatrix}
$$

(33)

Eliminate x_1 from rows r_1 and r_2, we have

$$
\begin{array}{c}
 \\
r_1 \\
r \\
r \\
r_4
\end{array}
\begin{array}{c}
\begin{array}{ccc}
x_1 & x_2 & 1
\end{array} \\
\left[
\begin{array}{ccc}
0 & 3/2 & -2 \\
0 & 1/2 & -2/3 \\
0 & 1 & 1 \\
0 & 1 & 0
\end{array}
\right]
\end{array}
*
\begin{bmatrix}
x_1 \\
x_2 \\
1
\end{bmatrix}
\geq
\begin{bmatrix}
0 \\
0 \\
0
\end{bmatrix}
$$

(34)

Normalize 2nd column for x_2, we have

$$
\begin{array}{c}
 \\
r_1 \\
r \\
r \\
r_4
\end{array}
\begin{array}{c}
\begin{array}{ccc}
x_1 & x_2 & 1
\end{array} \\
\left[
\begin{array}{ccc}
0 & 1 & -4/3 \\
0 & 1 & -4/3 \\
0 & 1 & 1 \\
0 & 1 & 0
\end{array}
\right]
\end{array}
*
\begin{bmatrix}
x_1 \\
x_2 \\
1
\end{bmatrix}
\geq
\begin{bmatrix}
0 \\
0 \\
0
\end{bmatrix}
$$

(35)

From (33), it is clear that x_2 is bounded below only with MLB=$\max\{-1,0,4/3\}=4/3$.

Also x_2 is not bounded above with LUB $= \infty$. Hence, the feasible interval for x_2 is $f_{x_2} = [4/3, \infty)$.

GGE FOR DIFFERENTIAL VARITIONAL INEQUALITIES (DVI)

GGE may also be applied to linear space as a normed Banach space with an inner-product operator with norm $\|u\| = \sqrt{<u, u>}$ for solving variational inequalities (VI) or differential variational inequalities (DVI) as follows [15 and 16] or Variational-like Inequalities in Banach space [25 to 29]:

Let $\phi = \{Bx \leq b\}$, where B is an $m \times n$ matrix as a nonempty convex compact polyhedron in R^n

Let F be a continuously differentiable function from ϕ into R^n with Jacobian F'.

The variational inequality problem (VIP) associated with F and ϕ is to locate a solution x^* in ϕ satisfying the variational inequality (VI): $(x^* - x)^T F(x^*) \leq 0 \forall x$ in ϕ. Note that in E^n, we have $\|x\| = \sqrt{<x, x>} = \sqrt{x^T x}$

Let the gap function associated with a VIP be defined for x in ϕ as:

$$g(x) = \max_{y \in \phi}(x - y)^T F(x)$$

While the dual gap function associated with a VIP is defined as $\bar{g}(x) = \max_{y \in \phi}(x - y)^T F(y)$

Using Newton's first order Taylor linear approximation around a point x_k in ϕ, a linearized VIP as LVIP can be computed iteratively for $k = 0, 1, 2, ...,$ as:

$$(x_{k+1} - y)^T (F(x_k) + F'(x_k)(x - x_k)) \leq 0 \forall y \in \phi.$$

Consider the following nonconvex, nonlinear constrained mathematical program:

$$\min_{x, \lambda} h(x, \lambda) \overset{def}{=} \lambda^T(b - Bx) = x^T F(x) + b^T \lambda \text{ subject to}$$
$$F(x) + B^T \lambda = 0, \ Bx \leq b, 0 \leq \lambda.$$

Note that optimality occurs at:

$$\min_{By \leq b} y^T F(x) = \max_{\substack{B^T \mu = F(x) \\ \mu \leq 0}} b^T \mu$$

subject to: $By \leq b, B^T \mu = F(x), \& \mu \leq 0$

Consequently, we have the following homogeneous linear feasible system of inequalities $f = Lw \geq 0$

$$L = \begin{vmatrix} B & 0 & b \\ 0 & B^T & -F(x) \\ 0 & -B^T & F(x) \\ 0 & -I & 0 \end{vmatrix} \text{ and } w = \begin{bmatrix} y \\ \mu \\ 1 \end{bmatrix}$$

Where

Note that $f = Lw \geq 0$ can be resolved effectively for the feasible intervals of w derived from L as described and demonstrated by the proposed GGE algorithm illustrated in Section 3.

CONCLUSIONS AND FUTURE WORK

In conclusion, the author proposes a generalization of the traditional Gaussian elimination (GE) for solving system of linear equalities to compute the feasible intervals of all variables to resolve the feasibility of all linear systems with both equalities and/or inequalities included. This Generalized Gaussian Elimination (GGE) for linear systems is applicable to a wide range of engineering and scientific applications and is related closely to the NPC mystery of the operations research and solvability of differential variation inequalities (DVI). Furthermore, it can be shown that GGE is indeed a special case of GE and that both GGE and GE do share the same worst case computational complexity of $O[n^2m]$ where n is the number of variables and m is the number of constraints. This is accomplished by replacing the variable substitution of the Gaussian elimination method by variable transition such that a specific variable may be safely and recursively eliminated without losing its binding inequalities and preserving both the most lower bounds (MLB) and the least upper bound (LUB). It is shown that any system of linear system with mixed linear equalities and inequalities may be converted into its standard homogeneous form such that the proposed GGE algorithm may be applied to obtain the feasible interval of any variable of choice. From the feasible intervals of all the variables of a given linear system, one may determine whether or not it contains binary, integers, or mixed solutions. The correctness and validity of GGE is illustrated by solving sample linear programs with unique solution, unbounded solution, and no solution.

Future work of this research includes the implementation of GGE as java code and Excel VB functions for very large system of linear inequalities or mixed of linear equalities and inequalities with millions of variables and/or constraints. The author is currently verifying a parameterized GGE algorithm for solving the linear Integer programming (LIP) problem as a potential alternative or replacement for the well known branch and bound (B&B) technique. A draft paper will be available for external review later. Funding from NSF or private foundations will be pursued to speed up the development of Java or VB functional codes for solving eigenvector systems, computing orthogonal basis, and DVI applications. GGE may also be applied to problems in operations research to reveal the availability of integer or binary solutions encountered in the NPC mystery as open issues.

ACKNOWLEDGEMENTS

Generalization of the classical Gaussian elimination for linear systems to cover inequalities is the result of years of inquiry and discussion with Dr. Leone C. Monticone and Dr. William P. Niedringhaus, colleagues of the author at the MITRE Corporation where the author served as an aviation engineer. The author also was benefited professionally and intellectually from his colleagues, Dr. Leonard Wojcik and Mr. Matt McMahon during many years at the MITRE Corporation working on MITRE sponsored research (MSR) projects related to solving vary large linear systems. Review and editorial advice from SAP and AJCAM are also essential to the improved quality and readability of this paper.

REFERENCES

1. Strang, Gilbert, "Introduction to Applied Mathematics", John Wiley & Sons Inc., New York, 1979.

2. Strang, Gilbert, "Karmarkar's algorithm and its place in applied mathematics", The Msthematical Intelligencer 9(2): pp. 4-10, New York: Springer, 1987.

3. Dantzig, G. G. "Maximization of a llinear function of variables subject to linear inequalities", 1947, Published pp. 339-347, in T.C. Koopmans (ed.): Activity Analysis of Production and Allocation, Wiley & Chapman-Hall, New York-Lodon, 1951.

4. Dantzig, G. B. "Linear Programming and Extensions". Princeton, NJ: Princeton University Press, 1963.

5. Fukuda, Komei and Terlaky, Tamas, "Crisis-cross methods: A fresh view on pivot algorithms", Mathematical Programming: Series B, No. 79, Papers from 16th International Symposium on Mathematical Programming, Lausanne, 1997.

6. Khachiyan, L. G., "Polynomial algorithms in linear programming", U.S.S.R., Computational Mathematical and Mathematical Physics 20 (1980) pp. 53-72.

7. Karmarkar, N., "A New Polynomial Time Algorithm for Linear Programming", AT&T Bell Laboratories, Murray Hill, New Jersey, September, 1984.

8. Gondzio, Jackek and Terlaky, Tamas, "A computational view of interior point method", Advances in linear and integer programming, Oxford Lecture Series in Mathematics and its Applications, 4, New York, Osford University Press. pp. 103-144, MR1438311, 1996.

9. Nocedal, Jorge and Wright, Stephen J,: "Numerical Optimization",

Springer Science+Business Media, Inc., 1999.

10. Michael. R. Garey and David. S. Johnson, COMPUTERS AND INTRACTABILITY, A Guide to the Theory of NP-Completeness, Bell Laboratories, Murray Hill, New Jersey, 1979.

11. Nemirovsky, A. and Yudin, N. "Interior-Point Polynomial Methods in Convex Programming", Philadelphia, PA: SIAM, 1994.

12. Alexander Schrijver, "Theory of Linear and Integer Programming", Department of Econometrics", Tilburg University, A Wiley-Interscience Publication, New York, 1979.

13. Niedringhaus, W., "Stream Option Manager (SOM): Automated Integration of Aircraft Separation, Merging, Stream Management, and Other Air Traffic Control Problems", IEEE Transactions Systems. Man & Cybernetics, Vol. 25 No. 9, Sept. 1995.

14. Niedringhaus, W., "Maneuver Option Manager (MOM): Automated Simplification of Complex Air Traffic Control Problems", IEEE Transactions Systems. Man & Cybernetics, May 1992. Niedringhaus, W., "Maneuver Option Manager (MOM): Automated Simplification of Complex Air Traffic Control Problems", IEEE Transactions Systems. Man & Cybernetics, May 1992.

15. Marcotte, Patrice, "A New Algorithm for Solving Variational Inequalties with Application to the Traffic Assignment Problem", Centre de Recherche sur les Transports, University de Montreal, Canada, Mathematical Programming 33 (1985) pp. 339-351, North-Holland.

16. Sun, Min, "A New Alternating Direction Method for Co-corecive Variational Inequality Problems with Linear Equality and Inequality Constraints", pp. 161-176, Advanced Modeling and Optimization, Vol. 12, Number 2, 2010.

17. Wang, Paul T. R., "Solving Linear Programming Problems in Self-dual Form with the Principle of Minmax", MITRE MP-89W00023, The MITRE Corporation, July, 1989.

18. Wang, Paul T. R., Niedringhaus, William P., and McMahon, Matthew T., "A Generic Linear Inequalities Solver (LIS) with an Application for Automated Air Traffic Control". America Journal of Computational and Applied Mathematics, pp. 195-206, Volume 3, Number 4, August 2013, http://article.sapub.org/10.5923.j.ajcam.20130304.01.html

19. Wang, Paul T. R., "Solving System of Linear Inequalities on the Surface of the Unit Shell (LIS-III)". American Journal of Computational and Applied Mathematics, pp. 97-110, Volume 3, Number 4, August 2014, http://article.sapub.org/10.5923.j.ajcam.20140403.05.html.

20. EE236-A, Lecture 15," Self-dual formulations", University of California, Department of Electrical Engineering, 2007-08.

21. Self-DualForm:http://www.ee.ucla.edu/ee236a/lectures/hsd.pdf.

22. ILOG, "Introduction to ILOG CPLEX", 2007, http://www.ilog.com/products/optimization/qa.cfm?presentation=3.

23. ILOG, "CPLEX Barrier Optimizer", 2008, http://www.ilog.com/products/cplex/product/barrier.cfm.

24. Steven Skiena, "LP_SOLVE: Linear Programming Code", Stony Brook University, Dept. of Computer Science", 2008 http://www.cs.sunysb.edu/~algorith/implement/lpsolve/implement.shtml.

25. Deepmala, "A Study on Fixed Point Theorems for Nonlinear Contractions and its Applications", Ph.D. Thesis (2014), Pt. Ravishankar Shukla University, Raipur (Chhatisgarh) India-492010.

26. V.N. Mishra, "Some Problems on Approximations in Banach Spaces", Ph.D. Thesis (2007), Indian Institute of Technology, Roorkee – 247 667, Uttarakhand, India.

27. S. Husain, S. Gupta, V.N. Mishra; "An Existence Theoram of Solutions for the System of Generalized Vector Quasi-Variational-Like Inequalities", American Journal of Operations Research (AJOR), 2013, 3.329-336. DOI:10.4236/AJOR.2013.22029.

28. S. Husain, S. Gupta, V.N. Mishra; "Generalized H(.,.)--Cocoercive Operators and Generalized Set-Valued Variational-Like Inclusions", Journal of Mathematics, Vol. 2013, Article3 ID 738491, 10 pages.

29. S. Husain, S. Gupta, V.N. Mishra; "Graph Convergence for the H((.,.),(.,.))-Mixed Mapping with an Application for Solving the System of Genberalized Variational Inclusions" Fixed Point Theory and Applications 2013, 2013:304 DOI: 10.1186/1687-1812-2013-304.

Chapter 8

THE ARITHMETIC MEAN SOLVER IN LAGGED DIFFUSIVITY METHOD FOR NONLINEAR DIFFUSION EQUATIONS

Emanuele Galligani

Department of Mathematics "G. Vitali", University of Modena and Reggio Emilia, Via Campi 213/b, I-41125, Modena, Italy

ABSTRACT

This paper deals with the solution of nonlinear system arising from finite difference discretization of nonlinear diffusion convection equations by the lagged diffusivity functional iteration method combined with different inner iterative solvers. The analysis of the whole procedure with the splitting methods of the Arithmetic Mean (AM) and of the Alternating Group Explicit (AGE) has been developed. A comparison in terms of number of iterations has been done with the BiCG-STAB algorithm. Some numerical experiments have been carried out and they seem to show the effectiveness of the lagged diffusivity procedure with the Arithmetic Mean method as inner solver.

INTRODUCTION

We consider a nonlinear diffusion convection equation where the diffusion coefficient, denoted by σ, depends on the solution.

When we use a finite difference discretization, this elliptic equation supplemented by a suitable boundary condition, can be transcribed into a nonlinear system of algebraic equations.

We wish to compute a solution of this system of nonlinear equations with a common iterative procedure in which the nonlinear term, corresponding to the discretization of the diffusivity σ, may be evaluated at the previous iteration (see[22]). In literature, this approach of *nonlinearity lagging* in the diffusivity term is denoted as Lagged Diffusivity Fixed Point Iteration or Lagged

Diffusivity Functional Iteration. In Section 2, a *model problem* described by a nonlinear diffusion convection equation subject to homogeneous Dirichlet boundary conditions is presented and a finite difference discretization is described. Then, the lagged diffusivity procedure for the solution of this nonlinear difference system is stated.

Since, a purpose here is to re-examine the lagged diffusivity procedure for solving the system of nonlinear difference equations of elliptic type in the context of Parallel Computing, the linear difference system that arises at each iteration of the lagged diffusivity procedure is solved with the iterative splitting methods of the Arithmetic Mean introduced in[20-21] and of the Alternating Group Explicit (AGE) introduced by Evans (see[1-3]).

Thus, the *outer* iterates of the lagged diffusivity procedure are the approximate solutions of the linear systems computed with an *inner* iterative solver; a criterion for acceptability of these approximate solutions is given. A stopping rule for the lagged diffusivity procedure is also given.

Section 3 is devoted to remind the Arithmetic Mean method and AGE method for the solution of linear systems with block tridiagonal coefficient matrix. In Section 4, the convergence of the lagged diffusivity iteration method is analysed under mild and reasonable assumptions imposed on the diffusivity σ using well known standard techniques (see[16]).

In Section 5, numerical experiments show the behaviour of the inner-outer iterations of the procedure. In this section a comparison of the lagged diffusivity iteration method with different iterative solvers is also presented. The Arithmetic Mean and the AGE methods are compared, in terms of number of iterations, with BiCG-STAB method (see[23]).

THE LAGGED DIFFUSIVITY FUNCTIONAL ITERATION METHOD

Consider the model problem described by the nonlinear diffusion convection equation

$$-div(\sigma\nabla u) + \boldsymbol{p}\cdot\nabla u + qu = f, \qquad (1)$$

where $u = u(x,y)$ is the density function at the point (x,y) of a diffusion medium R, $\sigma = \sigma(u) > 0$ is the diffusion coefficient or diffusivity and is dependent on the solution u, $q = q(x,y) \geq 0$ is the absorption term, the velocity vector $\boldsymbol{p} = (p_1, p_2)^T$ is assumed to be constant and $f(x,y)$ is a real valued sufficiently smooth function.

In the boundary ∂R of R, equation (1) can be supplemented by a homogeneous Dirichlet boundary condition of the form

$$u(x, y) = 0. \qquad (2)$$

In the following, we suppose R to be a rectangular domain with boundary ∂R and we assume that the functions σ, q and f satisfy the "smoothness" conditions:

(i) the function $\sigma = \sigma(u)$ is continuous in u; the functions $q(x,y)$ and $f(x,y)$ are continuous in x,y respectively;

(ii) there exist two positive constants σmin and σmax such that

$$0 < \sigma_{min} \leq \sigma(u) \leq \sigma_{max},$$

uniformly in u; in addition, $q(x,y) \geq qmin \geq 0$;

(iii) for fixed $(x,y) \in R$, the function $\sigma(u)$ satisfies Lipschitz condition in u with constant Γ (uniformly in x, y), $\Gamma > 0$.

The nonlinearity introduced by the u-dependence of the coefficient $\sigma(u)$ requires that, in general, the solution of equation (1) is approximated by numerical methods.

We superimpose on $R \cup \partial R$ a grid of points $R_h \cup \partial R_h$; the set of the internal points R_h of the grid are the mesh points (x_i, y_j), for $i = 1, ..., N$ and $j = 1, ..., M$, with uniform mesh size h along x and y directions respectively, i.e. $x_{i+1} = x_i + h$ and $y_{j+1} = y_j + h$ for $i = 0, ..., N, j = 0, ..., M$.

Thus, at the mesh points of $R \cup \partial R$, (x_i, y_j) for $i = 0, ..., N+1, j = 0, ..., M+1$, the solution $u(x_i, y_j)$ is approximated by a grid function u_{ij} defined on $R_h \cup \partial R_h$ and satisfying the boundary condition (2) on ∂R_h.

In order to approximate partial derivatives in (1) we shall make use of difference quotients of grid functions. The forward, backward and centered difference quotients with respect to x and to y of the grid function u_{ij} at the mesh point (x_i, y_j), are, respectively:

$$\Delta_x u_{ij} = \frac{u_{i+1j} - u_{ij}}{h}, \quad \Delta_y u_{ij} = \frac{u_{ij+1} - u_{ij}}{h},$$

$$\nabla_x u_{ij} = \frac{u_{ij} - u_{i-1j}}{h}, \quad \nabla_y u_{ij} = \frac{u_{ij} - u_{ij-1}}{h},$$

$$\delta_x u_{ij} = \frac{1}{2} (\Delta_x u_{ij} + \nabla_x u_{ij}),$$

$$\delta_y u_{ij} = \frac{1}{2} (\Delta_y u_{ij} + \nabla_y u_{ij}),$$

while the centered second difference quotient with respect to x and to y can be written

$$\delta_x^2 u_{ij} = \nabla_x \Delta_x u_{ij} = \Delta_x \nabla_x u_{ij},$$
$$\delta_y^2 u_{ij} = \nabla_y \Delta_y u_{ij} = \Delta_y \nabla_y u_{ij}.$$

This notation was introduced by Courant et al. in[4].

Providing a discretization error $O(h^2)$, the finite difference approximation of (1) in (x_i, y_j), $i = 1, ..., N, j = 1, ..., M$, is given by

$$-\Delta_x\left(\sigma(u_{ij})\nabla_x u_{ij}\right) - \Delta_y\left(\sigma(u_{ij})\nabla_y u_{ij}\right) +$$
$$+p_1 \delta_x u_{ij} + p_2 \delta_y u_{ij} + q(x_i, y_j)u_{ij} = f(x_i, y_j), \tag{3}$$

By ordering in a row lexicographic order the mesh points $P_l = (x_i, y_j)$, (i.e., $l = (j-1) \cdot N + i$ with $j = 1, ..., M$, and $i = 1, ..., N$), we can write the vector $\boldsymbol{u}$ of components u_{ij} and the difference equations (3) as the nonlinear system

$$F(\boldsymbol{u}) \equiv A(\boldsymbol{u})\boldsymbol{u} - \boldsymbol{f} = 0, \tag{4}$$

where the matrix $A(\boldsymbol{u})$ is of order $n = MN$ and has the block tridiagonal form; the M diagonal blocks are tridiagonal matrices of order N and the $M-1$ sub- and super- diagonal blocks are diagonal matrices of order N.

The five nonzero elements of $A(\boldsymbol{u})$ corresponding to u_{ij-1}, u_{i-1j}, u_{ij}, u_{i+1j} respectively, are $-(B_{ij} + \hat{B}_{ij})$, $-(L_{ij} + \hat{L}_{ij})$, $(D_{ij} + \hat{D}_{ij})$, $-(R_{ij} + \hat{R}_{ij})$, and $-(T_{ij} + \hat{T}_{ij})$, where

$$L_{ij} \equiv L_{ij}(\boldsymbol{u}) = \frac{1}{h^2}\sigma(u_{ij}), \quad R_{ij} \equiv R_{ij}(\boldsymbol{u}) = \frac{1}{h^2}\sigma(u_{i+1j}), \tag{5}$$

And

$$\hat{L}_{ij} = \frac{p_1}{2h}, \quad \hat{R}_{ij} = -\frac{p_1}{2h}, \quad \hat{B}_{ij} = \frac{p_2}{2h}, \quad \hat{T}_{ij} = -\frac{p_2}{2h},$$

$$D_{ij} \equiv D_{ij}(\boldsymbol{u}) = B_{ij} + L_{ij} + R_{ij} + T_{ij},$$
$$\hat{D}_{ij} = q(x_i, y_j). \tag{6}$$

The matrix $A(\boldsymbol{u})$ is an irreducible matrix ([24, p. 18]).

Providing that the mesh spacing h is sufficiently small, i.e.,

$$h < min\left\{\frac{2\sigma_{min}}{|p_1|}, \frac{2\sigma_{min}}{|p_2|}\right\}, \tag{7}$$

the matrix $A(\boldsymbol{u})$ is strictly (when $q(x_i, y_j) > 0$) or irreducibly (when $q(x_i, y_j) > 0$)

diagonally dominant ([24, p. 23]) and has positive diagonal elements, $a_{rr}(u) > 0$, $r = 1, ..., n$, and nonpositive off diagonal elements $a_{rs}(u) \leq 0$, $r \neq s$, with $r,s = 1$, ..., n; therefore $A(u)$ is an $\mathcal{M}$-matrix ([24, p. 91] or[18, p. 110]).

In the case of diffusion equation $(p = 0)$, the matrix $A(u)$ is also symmetric; then it is a symmetric and positive definite matrix ([24, p. 91]).

The vector f in (4) has components $f_i = f(x_i, y_j)$ for $i = 1, ..., N, j = 1, ..., M$ and $l = (j-1) \cdot N+i$.

We remark that while the grid function u_{ij} is defined on the whole mesh region $R_h \cup \partial R_h$, the vector $u \in \mathbb{R}^n$ represents the grid function $\{u_{ij}\}$ defined only on the interior mesh points R_h.

Here, we suppose that a solution u^* of the system (4) exists in $\mathcal{B}$ (see Section 4).

For solving the nonlinear system (4) the easiest and maybe the most common method is *to lag* the nonlinear term in (4) generating an iterative procedure denoted as Lagged Diffusivity Functional Iteration.

With this iterative procedure the nonlinear system (4) can be solved via a sequence of systems of linear equations.

Specifically, given a sequence of positive numbers $\{\varepsilon_v\}$ such that $\varepsilon_v \rightarrow 0$ as $v \rightarrow \infty$ and an initial estimate $u^{(0)}$ of the solution u^* of the system (4), we generate a sequence of iterates $\{u^{(v)}\}$, $v = 0, 1, ...,$ with the following rule for the transition from a current iteration $u^{(v)}$ to the new iterate $u^{(v+1)}$:

- Find an approximate solution $u^{(v+1)}$ of the linear system

$$F_v(u) \equiv A(u^{(v)})u - f = 0, \tag{8}$$

with the criterion for acceptability of the solution on the norm

$$\left\| F_v(u^{(v+1)}) \right\| \leq \varepsilon_{v+1}. \tag{9}$$

Then, the lagged diffusivity procedure is composed by an *outer* iteration that generates the sequence $\{u^{(v)}\}$, and by an *inner* iterative solver of the linear system (8). This solver must be particularly well suited for implementation on parallel computers.

The termination criterion for the outer iteration is provided by the following stopping rule

$$\varepsilon_{v+1} \leq \varepsilon,$$

(10)

where ε_{v+1} decreases as $\varepsilon_{v+1} = 0.5\ \varepsilon_v$, $v = 1,2,....$, with $\varepsilon_1 = 0.1\|F(u^{(0)})\|$ and ε is a prespecified threshold.

ITERATIVE PARALLEL SOLUTION OF THE LINEAR SYSTEMS

In this section, we remind the block form of the Alternating Group Explicit (AGE) and of the Arithmetic Mean methods for the solution of the linear system

$$Ax = b,$$

(11)

when the nn matrix A has the block tridiagonal form

$$A = \begin{pmatrix} B_1 & C_1 & & & & \\ A_2 & B_2 & C_2 & & & \\ & A_3 & B_3 & C_3 & & \\ & & \ddots & \ddots & \ddots & \\ & & & A_{M-1} & B_{M-1} & C_{M-1} \\ & & & & A_M & B_M \end{pmatrix}$$

(12)

Here, each square block B_i, $i = 1, ..., M$, is a nonsingular NN matrix and the blocks A_i and C_i $(i = 1, ..., M;\ A_1 = C_M = 0)$ are square matrices of order $N\ (n = NM)$.

The AGE method consists in considering the following splitting of the matrix A $A = G_1 + G_2$, where G_1 and G_2 are the following matrices

$$G_1 = \begin{pmatrix} B_1 & C_1 & & & & & \\ A_2 & B_2' & & & & & \\ & & B_3' & C_3 & & & \\ & & A_4 & B_4' & & & \\ & & & & \ddots & & \\ & & & & & B_{M-1}' & C_{M-1} \\ & & & & & A_M & B_M' \end{pmatrix},$$

$$G_2 = \begin{pmatrix} B_1' & & & & & & \\ & B_2' & C_2 & & & & \\ & A_3 & B_3' & & & & \\ & & & \ddots & & & \\ & & & & B_{M-2}' & C_{M-2} & \\ & & & & A_{M-1} & B_{M-1}' & \\ & & & & & & B_M' \end{pmatrix},$$

With $B_i' = \frac{1}{2}B_i, i = 1, ..., M.$

Thus, starting from a vector $x^{(0)}$, the AGE method generates a sequence of iterates $\{x^{(k)}\}$ as follows; for $r \geq 0$ and $k = 0, 1, ...$, until convergence:

$$(G_1 + rI_n)x^{(k+1/2)} = (rI_n - G_2)x^{(k)} + b,$$
$$(G_2 + rI_n)x^{(k+1)} = (rI_n - G_1)x^{(k+1/2)} + b.$$
$$(13)$$

Here, I_n is the identity matrix of order n. The AGE method is convergent when the matrices G_1 and G_2 are symmetric and positive definite with $r \geq 0$. In this case it is proved that the optimal choice for r is $r = \sqrt{ab}$ where

$$0 < a \leq \lambda(G_1) \leq b, \qquad 0 < a \leq \lambda(G_2) \leq b,$$

and $\lambda(G_s)$, $s = 1, 2$, are the eigenvalues of the matrix G_s (see, e.g.[1]).

Furthermore, if the matrix A is irreducibly (or strictly) diagonally dominant with positive diagonal elements, $a_{ii} > 0$, and nonpositive off diagonal elements $a_{ij} \leq 0, i \neq j$, and

$$r > \max_{1 \leq i \leq n} \frac{a_{ii}}{2}, \quad \left(or \quad r \geq \max_{1 \leq i \leq n} \frac{a_{ii}}{2} \right),$$
$$(14)$$

then the AGE method is convergent.

Indeed, from the hypotheses we have that the matrix A is an M-matrix. Set $P_s = G_s + rI_n$ and $Q_s = rI_n - G_s$, $s = 1, 2$. The choice of r in (14) yields that the matrices P_s are strictly (or irreducibly) diagonally dominant and have positive diagonal elements and nonpositive off diagonal elements, then P_s are M-matrices with $P_s^{-1} \geq 0$. From the hypotheses on the matrix A, the matrices Q_s are nonnegative $(Q_s \geq 0, s = 1, 2)$

Thus, the iteration matrix T of the AGE method is

$$T = P_2^{-1}Q_1P_1^{-1}Q_2 \geq 0.$$

Set

$$P^{-1} = 2r\,P_2^{-1}P_1^{-1},$$
$$(15)$$

we have $P^{-1} \geq 0$ (P is an M-matrix) and $T = I_n - P^{-1}A$ holds.

Indeed, set $A = P_1 - Q_2 = P_2 - Q_1$, a multiplicative splitting method can be written

$$P_1 x^{(k+1/2)} = (P_1 - A)x^{(k)} + b,$$
$$P_2 x^{(k+1)} = (P_2 - A)x^{(k+1/2)} + b.$$

Then,

$$x^{(k+1)} = (I_n - P_2^{-1}A)(I_n - P_1^{-1}A)x^{(k)}$$
$$+ ((I_n - P_2^{-1}A)P_1^{-1} + P_2^{-1})\,b,$$

can be seen as a splitting method (P, Q), i.e., $A = P - Q$, by setting

$$P^{-1} = ((I_n - P_2^{-1}A)P_1^{-1} + P_2^{-1}) = P_2^{-1}(P_1 + P_2 - A)P_1^{-1}.$$

Since $P_s = G_s + rI_{n^2}$, $s = 1, 2$, and $A = G_1 + G_2$, then we have the expression (15) for P^{-1}.

Now, the proof runs as that of the Regular Splitting Theorem in[18, p. 119].

The Arithmetic Mean method uses the following two splittings of the matrix A

$$A = H_1 + K_1, \quad A = H_2 + K_2,$$

where H_1 and H_2 are the following matrices

$$H_1 = \begin{pmatrix} B_1 & C_1 & & & & & \\ A_2 & B_2 & & & & & \\ & & B_3 & C_3 & & & \\ & & A_4 & B_4 & & & \\ & & & & \ddots & & \\ & & & & & B_{M-1} & C_{M-1} \\ & & & & & A_M & B_M \end{pmatrix},$$

$$H_2 = \begin{pmatrix} B_1 & & & & & \\ & B_2 & C_2 & & & \\ & A_3 & B_3 & & & \\ & & & \ddots & & \\ & & & & B_{M-2} & C_{M-2} \\ & & & & A_{M-1} & B_{M-1} \\ & & & & & B_M \end{pmatrix},$$

and $K_1 = A - H_1$, $K_2 = A - H_2$.

We suppose M even. If M is odd, we can proceed in a similar way.

Thus, starting from a vector $x^{(0)}$, the method of the Arithmetic Mean generates a sequence of iterates $\{x^{(k)}\}$ as follows; for $\rho \geq 0$ and $k = 0, 1, ...,$ until convergence:

$$(H_1 + \rho I_n)z^{[1]} = (\rho I_n - K_1)x^{(k)} + b,$$
$$(H_2 + \rho I_n)z^{[2]} = (\rho I_n - K_2)x^{(k)} + b,$$
$$x^{(k+1)} = \frac{1}{2}(z^{[1]} + z^{[2]}).$$
(16)

In the paper [21], it is proved that the block form of the Arithmetic Mean method above described, is convergent when the matrix A is:

- irreducibly (strictly) diagonally dominant with positive diagonal entries and nonpositive off diagonal elements with $\rho > 0$ ($\rho \geq 0$);

- positive definite but not symmetric with $\rho > \rho^*$ where

$$\rho^* = \max\{\frac{|\lambda_{min}(F_1)|}{|\lambda_{min}(\tilde{A})|}, \frac{|\lambda_{min}(F_2)|}{|\lambda_{min}(\tilde{A})|}\}$$

Here $\lambda(V)$ denotes an eigenvalue of a matrix V and

$$F_s = H_s H_s^T - K_s K_s^T, \qquad s=1,2,$$

and $\tilde{A}$ is the symmetric positive definite matrix $\tilde{A} = A + A^T$;

- symmetric positive definite with $\rho \geq 0$.

At each iteration k of the AGE method, we have to solve $M/2$ linear systems of order $2N$, $i = 1, 3, 5, ..., M-1$,

$$\begin{bmatrix} B_i' + rI_N & C_i \\ A_{i+1} & B_{i+1}' + rI_N \end{bmatrix}\begin{bmatrix} x_i \\ x_{i+1} \end{bmatrix} = \begin{bmatrix} v_i \\ v_{i+1} \end{bmatrix},$$
(17)

to obtain the vector $x^{(k+1/2)}$. The solution of (17) can be seen as a block partitioned vector

$$x^{(k+1/2)} = (x_1^{\left(k+\frac{1}{2}\right)^T}, x_2^{\left(k+\frac{1}{2}\right)^T}, ..., x_M^{\left(k+\frac{1}{2}\right)^T})^T,$$ where each block has N components.

Here $v = (v_1^T, v_2^T, ..., v_M^T)^T$ is the right hand side (r.h.s.) of the first equation of (13).

These $M/2$ systems can be solved simultaneously (*in parallel*).

Then, in order to obtain the new iterate $x^{(k+1)}$ of the AGE method, we have to solve, in parallel, $M/2-1$ systems as (17) with $i = 2, 4, 6, ..., M-2$ and two linear systems of order N

$$\left(B_1' + rI_N\right)x_1 = v_1,$$
$$\left(B_M' + rI_N\right)x_M = v_M, \text{ that can be solved in parallel, as well. The AGE method}$$
has an intrinsic parallelism.

In the case of the additive splitting method of the Arithmetic Mean, at each iteration k we have to solve M-1 linear systems of order $2N$, $i = 1, 2, 3, ..., M$-1,

$$\begin{bmatrix} B_i + \rho I_N & C_i \\ A_{i+1} & B_{i+1} + \rho I_N \end{bmatrix} \begin{bmatrix} z_i \\ z_{i+1} \end{bmatrix} = \begin{bmatrix} w_i \\ w_{i+1} \end{bmatrix}, \tag{18}$$

where $z = (z_1^T, z_2^T, ..., z_M^T)^T$ and $w = (w_1^T, w_2^T, ..., w_M^T)^T$ indicate the vector $z^{[1]}$ (for $i = 1, 3, 5,$ ..., M-1) and the corresponding r.h.s. of the first equation of (16) or the vector $z^{[2]}$ (for $i = 2, 4, 6, ..., M$-2) and the corresponding r.h.s. of the second equation of (16).

Furthermore, we have to solve two linear systems of order N

$$\left(B_1 + \rho I_N\right)z_1 = w_1$$
$$\left(B_M + \rho I_N\right)z_M = w_M$$

where z_1 and z_M indicate the first and the last block of the vector $z^{[2]}$ and w_1, w_M the corresponding r.h.s. of the second equation of (16).

These systems can be solved in parallel. The Arithmetic Mean method introduces also an *explicit parallelism* in order to increase the degree of multiprogramming, that is the number of processes that can be executed simultaneously ([13, p. 87]).

When the system (11) arises from the finite difference discretization of the problem (1)-(2), the diagonal blocks of A in (12) are tridiagonal while the sub and superdiagonal blocks are diagonal.

Thus, the systems (17) and (18) can be solved directly as in[9, 8] or iteratively, generating a two-stage iterative method as in[12]. A direct solution of these systems can be performed by cyclic reduction solvers ([17, p. 125],[15]) combined with an approximate Schur complement method ([17, p.123, p. 217]) or with block Gaussian elimination.

Analysis of the Convergence of the Lagged Diffusivity Functional Iteration Method

In this section, we prove the convergence of the sequence $\{u^{(\nu)}\}$ generated by the lagged diffusivity iteration method for the solution of the system (4) under the smoothness assumptions (i)-(iii). We define u_{ij} and v_{ij}, $i = 0, ..., N$ +1 and j = 0, ..., M +1, the grid functions defined on $R_h \cup \partial R_h$ and satisfying the Dirichlet boundary condition on ∂R_h. For grid functions $\{u_{ij}\}$ and $\{v_{ij}\}$ of this type, the discrete $l_2(R_h)$ inner product and norm are defined by the formulas

$$\langle u, v \rangle = h^2 \sum_{i=1}^{N} \sum_{j=1}^{M} u_{ij} v_{ij},$$

$$\| u \|_h = \left(h^2 \sum_{i=1}^{N} \sum_{j=1}^{M} |u_{ij}|^2 \right)^{1/2} = \langle u, u \rangle^{1/2},$$

respectively.

We say that the grid functions $\{u_{ij}\}$ defined on $R_h \cup \partial R_h$ and vanishing on ∂R_h satisfy **Property A** if they are uniformly bounded and have uniformly bounded backward difference quotients $\nabla_x u_{ij}$ and $\nabla_y u_{ij}$ at each mesh point (x_i, y_j) of $R_h \cup \partial R_h$. The set of all grid functions which satisfy Property A is denoted by $\mathcal{B}$. Thus, $\mathcal{B}$ is the set of all grid functions $\{u_{ij}\}$ for which there exist two positive constants ϱ and such that

$$\| u \|_h \leq \varrho, \tag{19}$$

$$|\nabla_x u_{ij}| \leq \beta, \qquad |\nabla_y u_{ij}| \leq \beta, \tag{20}$$

The constant ϱ is independent of h; also β is independent of h but it depends on $\| f \|$.(see[16]).

We assume that the system (4), $F(u) = 0$, where $F(u) \equiv A(u) u - f$, has at least one solution $u^* \in \mathcal{B}$.

Suppose that the system (4) arises from the discretization of problem (1)-(2) subject to the conditions (i)-(iii) with $q \geq q_{min} > 0$ and $A(u)$ being an irreducible nonsingular M-matrix. If

$$\frac{\Gamma^2 \beta^2}{2 q_{min} \sigma_{min}} < 1$$

for any $u \in \mathcal{B}$, then the mapping $F(u)$ is uniformly monotone in $\mathcal{B}$. (See[16],[5] for a proof of this result and e.g.,[19, p. 141] for the definition of uniform monotonicity).

The iterate $u^{(v+1)}$ of the lagged diffusivity functional iteration method satisfies the system (8) with the acceptability criterion (9), that is $u^{(v+1)}$ is the solution of the linear diffusion convection equation whose diffusivity depends on the previous iterate $u^{(v)}$ with inhomogeneous term $-f - F_v(u^{(v+1)})$.

for $i = 1, ..., N+1$ and $j = 1, ..., M+1$.

Thus, we can state the following result concerning the convergence of the lagged diffusivity functional iteration method.

Theorem 1. Let u^* be a solution of the nonlinear system (4) with $F(u) \equiv A(u)u - f$ arising from the finite difference discretization of problem (1)-(2) subject to the conditions (i)-(iii) with $q \geq q_{min} > 0$ and $A(u)$ being an irreducible nonsingular $\mathcal{M}$-matrix.

Assume that the mapping $F(u)$ is uniformly monotone in $\mathcal{B}$.

Suppose that $\{\varepsilon_\nu\}$ is a sequence of positive numbers such that $\varepsilon_\nu \to 0$ as $\nu \to \infty$.

Let $u^{(0)} \in \mathcal{B}$ be arbitrary and let $u^{(\nu+1)}$ be the solution of $F_\nu(u) = 0$ satisfying the condition (9) with $F_\nu(u)$ as in (8).

If all the vectors $u^{(\nu)}$ belong to $\mathcal{B}$ and satisfy Property A with (21) instead of (20), then the sequence $\{u^{(\nu)}\}$ converges to u^*.

Proof. The proof runs as that in[6, Theor. 1] or in[5, Theor. 1, p. 33].

A more general result on the convergence of the lagged diffusivity functional iteration method can be obtained by defining the mapping $u=G(u)$ where

$G(u) = A(u)^{-1}f$, for all $u \in \mathbb{R}^n$. A solution of the system (4) is a fixed point of the mapping $G(u)$.

Since the smoothness condition (iii) on $\sigma(u)$, we can have that the matrix $A(u)$ satisfies the Lipschitz-continuity condition for every bounded subset Ω of $\mathbb{R}^n$ with a Lipschitz constant Λ. Then we can write,

$G(u) - G(v) = A(u)^{-1}\big(A(v) - A(u)\big)A(v)^{-1}f$.

Thus, we have for $u, v \in \Omega$

$$\| G(u) - G(v) \| \leq \Lambda \| f \| \| A(u)^{-1} \| \times$$
$$\times\| A(v)^{-1} \| \| u - v \| \leq \Lambda \beta^2 \| f \| \| u - v \|$$

where β is a bound for $\| A(u)^{-1} \|$ for any $u \in \Omega$. Here, $\| \cdot \|$ indicates an arbitrary vector and matrix norm.

The last inequality assures that the mapping $G(u)$ satisfies a Lipschitz condition on a bounded subset Ω of $\mathbb{R}^n$.

NUMERICAL EXPERIMENTS

In this section, we consider a numerical experimentation of the lagged diffusivity functional iteration method for the solution of the nonlinear system (4) generated by the finite difference discretization above described, of the elliptic problem (1)-(2). Indeed, we have to solve the system

$A(u)u = f$.

In these experiments, the vector solution u^* is prefixed and it is composed by the values of the prescribed function $u(x, y) = \sin(\pi x)\sin(\pi y)$ defined on the square $[0, 1] \times [0, 1]$.

The chosen functions for $\sigma(u)$ are

$$\sigma 1 : \sigma(u) = 1 + u;$$
$$\sigma 2 : \sigma(u) = \frac{a}{b + c\,u};$$
$$\sigma 3 : \sigma(u) = 2(1 + 2u - 2u^2);$$

where, in the case of $\sigma 2$, we have $\sigma 2\_1$ for $a=3$, $b=2$, $c=1$; $\sigma 2\_2$ for $a=1.5$, $b=0.1$, $c=0.9$ and $\sigma 2\_3$ for $a=1$, $b=0.01$, $c=0.99$.

The vector f is computed as

$$f \equiv f^* = A(u^*)u^*,$$

where the matrix $A(u^*)$ of order n, has elements as in (5) and (6) with $N=M$ and $n = N^2$. In all the experiments we have $N=256$ and $p_1 = p_2 \equiv p$.

At each iteration v, $v = 0,1,...,$ of the lagged diffusivity procedure, we have to solve the linear system of order n $A(u^{(v)})u = f$, with the splitting method of the Arithmetic Mean or of the Alternating Group Explicit described in a previous section (see e.g.[9],[10] for an evaluation of the block form of the Arithmetic Mean method on different parallel architectures and[7] for a description of the Fortran code implementing the method). These methods are compared with BiCG-STAB method implemented as in[14, p. 50].

We call $u^{(v+1)}$ the new iteration of the lagged diffusivity procedure, computed with k_{v+1} iterations of the inner solver such that the inner residual

$$F_v(u^{(v+1)}) \equiv A(u^{(v)})u^{(v+1)} - f^*,$$ satisfies the condition (9)

$$\left\| F_v\left(u^{(v+1)}\right) \right\| \le \varepsilon_{v+1},$$ with $\varepsilon_1 = 0.1 \; \| F(u^{(0)}) \|$ and

$$\varepsilon_{v+1} = 0.5\,\varepsilon_v. \tag{22}$$

Here, $\|\cdot\|$ indicates the Euclidean norm.

The vector $F(u^{(0)}) = A(u^{(0)})u^{(0)} - f$ is the initial *outer residual* and its Euclidean norm is called $res0$.

The initial vector $u^{(0)}$ is taken as the null vector $(u^{(0)} = 0)$ or as the vector e whose all the components are equal to 1 $(u^{(0)} = e)$.

The lagged diffusivity procedure has been implemented in a Fortran code with machine precision 2.2×10^{-16} and stops when (10) holds, i.e., for $v = 0,1,...,$

$$\varepsilon_{v+1} \le \underline{\varepsilon},$$

with $\varepsilon = 10^{-4}$.

We call v^* the iteration of the lagged diffusivity procedure for which condition (10) is satisfied. That is, the iterate $u^{(v^*)}$ satisfies $\|F_{v^*-1}(u^{(v^*)})\| \leq \varepsilon_{v^*}$ $(\varepsilon_{v^*} > \varepsilon)$, (and we have $\varepsilon_{v^*+1} = 0.5 \, \varepsilon_{v^*}$ and $\varepsilon_{v^*+1} \leq \varepsilon$.

In the tables, we report the number of iterations v^* and, in brackets, the total number of iterations of the inner solver k_T, i.e.,

$$k_T = \sum_{v=1}^{v^*} k_v.$$

In the tables, we also report the discrete $l_2(R_h)$ norm of the error, $err = \| u^* - u^{v^*} \|_h$, the Euclidean norm of the outer residual

$$res = \left\| F\left(u^{(v^*)}\right) \right\| = \left\| A\left(u^{(v^*)}\right)u^{(v^*)} - f^* \right\|, \quad \text{and } res0.$$

The symbol * close to the value of *res* indicates that the behaviour of the norm of the outer residual $\|F_v(u^{(v+1)})\|$ is not monotone decreasing.

The writing *max it.* indicates that at a certain iteration v, the maximum number of iterations of the inner solver has been reached. The maximum number of inner iterations is set equal to 20000.

The writing *n.c.* indicates that at a certain iteration v, the condition (7) is not satisfied.

The symbol " ü « close to the number of inner and outer iterations denotes that at a certain iteration v, the condition (7) is not satisfied and, in these cases, the lagged diffusivity iteration method generates the iteration by performing a prefixed number (equal to 20) of iterations of the inner iterative solver.

The writing 1.71(-8) indicates $1.71 \cdot 10^{-8}$.

In the tables, we indicate with *lag-AM*, *lag-AGE* and *lag-B*, the lagged diffusivity functional iteration method with the Arithmetic Mean, the Alternating Group Explicit and the BiCG-STAB method, respectively, as inner solver.

Furthermore, we observe that, since ε_{v+1} decreases, for v increasing, as (22) and the lagged diffusivity functional iteration method stops at the iteration v^* when the criterion for ε_{v^*+1} in (10) is satisfied, we have

$$\varepsilon_{v^*+1} = \frac{1}{2}\varepsilon_{v^*} = \frac{1}{2}\varepsilon_{v^*-1} \cdots = \frac{1}{2^{v^*}}\varepsilon_1 \leq \varepsilon,$$

where we set $\varepsilon_1 = 0.1 \parallel \bar{F}\left(u^{(0)}\right) \parallel$. Then,

$$v^* > \log_2\left(\frac{\varepsilon_1}{\varepsilon}\right)$$

Indeed, in the experiments we obtain

$$v^* = \lceil \log_2\left(\frac{\varepsilon_1}{\varepsilon}\right) \rceil.$$

Table 1: Results for different values of p for $\sigma 1$

$N=256$	$\sigma(u) = \sigma 1$	$q=0$	$u^{(0)} = 0$	
p		$v^*(k_T)$		$res0$
	lag-AM	*lag-AGE*	*lag-B*	
500	29 (478) ✓	29 (538) ✓	29 (1399)	284368.27
400	28 (547)	28 (621)	28 (1189)	227507.49
300	28 (720)	28 (851)	28 (1118)	170650.88
200	27 (1062)	27 (1243)	27 (1175)*	113804.73
100	26 (2267)	26 (2709)	26 (1022)*	57000.26
50	25 (5131)	25 (6088)	25 (946)	28691.16
10	23 (31110)	23 (35322)	23 (1202)	6833.42
1	22 (55159)	22 (64165)	22 (1357)*	3820.91

Table 2: Results for different values of p for $\sigma 2\_1$

$N=256$	$\sigma(u) = \sigma 2\_1$	$q=0$	$u^{(0)} = 0$	
p		$v^*(k_T)$		$res0$
	lag-AM	*lag-AGE*	*lag-B*	
500	29 (441)	29 (470)	29 (1189)	284348.29
400	28 (524)	28 (565)	28 (1110)	227486.24
300	28 (681)	28 (757)	28 (1108)*	170627.52
200	27 (972)	27 (1088)	27 (1050)	113777.13
100	26 (1965)	26 (2208)	26 (1034)	56960.00
50	25 (4232)	25 (4857)	25 (864)	28625.88
10	23 (24409)	23 (30782)	23 (1149)	6605.37
1	22 (60546)	22 (79276)	22 (1581)	3418.60

Table 3: Results for different values of p and $^{(0)}$ for $\sigma2\_2$

	N=256	$\sigma(u) = \sigma2\_2$		$q=0$
p		$v^*(k_T)$		$res0$
	lag-AM	lag-AGE	lag-B	
		$u^{(0)} = 0$		
200	n.c.	n.c.	max it.*	121921.87
100	n.c.	n.c.	max it.*	72982.83
50	26 (21137)	26 (70137)	26 (2767)	54613.78
10	26 (56063)	26 (343861)	26 (3907)	47663.32
1	26 (80638)	max it.	26 (4449)	47481.15
		$u^{(0)} = e$		
500	35 (1272)	35 (4657) ✓	max it.*	21347550.45
400	35 (1616)	35 (5671)	max it.*	21581543.63
300	35 (2154)	35 (7633)	max it*	21820683.14
200	35 (3120)	35 (11226)	max it*	22064801.66
100	35 (6153)	35 (26619)	35 (2782)	22313735.75
50	35 (12048)	35 (57421)	35 (2724)	22439958.69
10	35 (46793)	35 (342113)	35 (4235)	22541761.97
1	35 (72524)	max it.	35 (4842)	22564767.64

Table 4: Results for different values of p and $^{(0)}$ for $\sigma2\_3$

	N=256	$\sigma(u) = \sigma2\_3$		$q=0$
p		$v^*(k_T)$		$res0$
	lag-AM	lag-AGE	lag-B	
		$u^{(0)} = 0$		
200	n.c.	n.c.	max it.*	746998.50
100	n.c.	n.c.	max it.*	744863.51
50	30 (23198)	max it.	max it.*	745422.36
10	30 (53247)	max it.	30 (10173)	746649.16
1	30 (102915)	max it.	30 (13616)	747020.37
		$u^{(0)} = e$		
500	38 (2486)* ✓	n.c.	max it.*	147793693.53
400	38 (2516) ✓	n.c.	max it.	148074183.84
300	38 (3108) ✓	38 (68052)	max it.*	148355269.76
200	38 (4255)	38 (108203)	max it.*	148636947.90
100	38 (7153)	max it.	max it.*	148919214.91
50	38 (11826)	max it.	max it.*	149060568.19
10	38 (40302)	max it.	max it.*	149173755.93
1	38 (93196)	max it.	max it.*	149199236.03

Table 5: Results for different values of p for $\sigma3$

$N=256$	$\sigma(u) = \sigma3$	$q=0$	$u^{(0)} = 0$	
p		$v^*(k_T)$		$res0$
	$lag\text{-}AM$	$lag\text{-}AGE$	$lag\text{-}B$	
500	29 (765)	29 (818)	29 (1161)*	284482.74
400	28 (953)	28 (1026)	28 (1081)	227640.62
300	28 (1253)	28 (1341)	28 (1039) *	170815.09
200	27 (1974)	27 (2118)	27 (1016)	114030.97
100	26 (4339)	26 (4683)	26 (924)	57411.25
50	25 (9725)	25 (10525)	25 (1054)	29460.86
10	24 (46400)	24 (53706)	24 (1565)	9468.12
1	23 (61128)	23 (71699)	$max\ it.$*	7559.22

Table 6: Results for different values of q

N=256	$\sigma(u) = \sigma1$		$u^{(0)} = 0$	
q		$v^*(k_T)$		$res0$
	$lag\text{-}AM$	$lag\text{-}AGE$	$lag\text{-}B$	
		$p = 20$		
0	24 (14136)	24 (18027)	24 (1069)	11991.73
10	24 (13217)	24 (16868)	24 (998)	12422.56
100	25 (8814)	25 (10794)	25 (600)	19939.78
1000	27 (2118)	27 (2383)	27 (211)	132449.08
		$p = 200$		
0	27 (1062)	27 (1243)	27 (1175)	113804.73
10	27 (1061)	27 (1245)	27 (1185)*	113850.93
100	27 (1039)	27 (1232)	27 (1124)*	114914.35
1000	28 (776)	28 (928)	28 (860)	174213.87
10000	31 (267)	31 (290)	31 (128)	1293461.86
N=256	$\sigma(u) = \sigma2\_3$		$p=100$	$u^{(0)} = e$
0	38 (7153)	$max\ it.$	$max\ it.$*	1489192 14.91
100	38 (6468)	38 (214471)	$max\ it.$*	148921465.88
1000	38 (3664)	38 (97128)	$max\ it.$*	148941809.24
10000	38 (1117)	38 (15122)	$max\ it.$*	149153606.71

Table 7: Some results for the error and the residuals

N=256	$\sigma(u) = \sigma 1$ 0	q=0	$u^{(0)} =$
method	err	res	res0
p=1			
lag-AM	1.71(-8)	1.54(-4)	3820.91
lag-AGE	1.71(-8)	1.62(-4)	
lag-B	7.78(-9)	1.90(-4)	
p=50			
lag-AM	1.49(-9)	1.69(-4)	28691.16
lag-AGE	9.71(-10)	1.71(-4)	
lag-B	1.62(-10)	1.70(-4)	
p=100			
lag-AM	5.03(-10)	1.69(-4)	57000.26
lag-AGE	3.12(-10)	1.69(-4)	
lag-B	4.30(-11)	4.90(-5)	
p=300			
lag-AM	6.89(-11)	1.19(-4)	170650.88
lag-AGE	4.24(-11)	1.20(-4)	
lag-B	1.42(-11)	1.21(-4)	

CONCLUSIONS

From the numerical experiments the following conclusions can be drawn:

1.   the outer residual **res** has the same order of ε and the error **err** in the discrete $l_2(R_h)$ norm has, in worst cases, order $h\varepsilon$;

2.   the AM method gives better results when the ratio between the maximum value of σ and the smallest component of p is small, that is the coefficient matrix of the linear system is strongly asymmetric (see[20]) or the deviation from asymmetry is decreasing. We define as the deviation from asymmetry of a matrix the difference between the Frobenius norms of the symmetric and nonsymmetric parts of the matrix. Furthermore, we can observe the same behaviour of the AGE method with the one of the AM method, respect to the deviation from asymmetry of the coefficient matrix that occurs at each step of the lagged diffusivity procedure;

3.   the lagged diffusivity functional iteration method combined with the AM or the AGE method breaks down when the coefficient matrix, at a certain iteration v, is not an M-matrix (n.c.). In some cases, the AGE inner solver, requires a large number of inner iterations especially for nearly symmetric matrices;

4. the behaviour of the residual *res* when we use AM or AGE method as inner solver, is always monotone, except in one case where the lagged diffusivity procedure breaks down (the coefficient matrix is not an $\mathcal{M}$ -matrix) but it has been possible to force the convergence by running a few number of iterations of the inner iterative solver. The nonmonotonicity of the residual happens at these "forced" iterations. This technique of forcing convergence is successful when condition (7) is "almost satisfied".

5. the lagged diffusivity functional iteration method combined with the BiCG-STAB method, when it does not break down, requires a number of inner iterations that is not large and seems to be independent from the deviation from asymmetry of the coefficient matrix. We observe that the failure of the lagged diffusivity procedure with the BiCG-STAB method, in most cases, happens when the decreasing of the residual *res* is not monotone;

6. the behaviour of the lagged diffusivity procedure with AM or AGE method as iterative solver depends on the choice of the initial vector. The choice $u^{(0)} = 0$ in the cases $\sigma(u) = \sigma2\_2$ or $\sigma(u) = \sigma2\_3$ yields to negative values for $u^{(v)}$ for a certain v (tipically $v = 1$ or $v = 2$) so that condition (7) is not satisfied; that is the initial iterate is too close to a region where a sufficient condition for determining the iterates of the outer procedure is not satisfied. Then, the control on condition (7) is a detection to start from another initial vector.

REFERENCES

7. D.J. Evans, "Group explicit iterative methods for solving large linear systems", Taylor & Francis, International Journal of Computer Mathematics, vol. 17, pp. 81-108, 1985.

8. D.J. Evans, W.S. Yousif, "The block alternating group explicit method (BLAGE) for the solution of elliptic difference equations", Taylor & Francis, International Journal of Computer Mathematics, vol. 22, pp. 177-185, 1987.

9. D.J. Evans, W.S. Yousif, "The solution of two-point boundary value problems by the alternating group explicit (AGE) method", SIAM Journal on Scientific and Statistic Computing, vol. 9, pp. 474-484, 1988.

10. R. Courant, K.O. Friedrichs, H. Lewy, Über die partiellen Differenzengleichungen der mathematischen Physik»», Springer, Mathematische Annalen, vol. 100, pp. 32-74, 1928.

11. [5]   E. Galligani, «A note on the iterative solution of nonlinear steady state reaction diffusion problems», University of Modena and

Reggio Emilia, Technical Report of Numerical Analysis TR NA-UniMoRE-1-2010, August 2010.

12. E. Galligani, «Lagged diffusivity fixed point iteration for solving steady state reaction diffusion problems», Taylor & Francis, International Journal of Computer Mathematics, in press 2012. (Available online: 2 April 2012. DOI:10.1080/00207160.2012.671478)

13. E. Galligani, «Analysis of the Fortran routines implementing the arithmetic mean method for the solution of block tridiagonal linear systems», University of Modena and Reggio Emilia, Technical Report of Numerical Analysis TR NA-UniMoRE-4-2012, March 2012.

14. E. Galligani, V. Ruggiero, «A parallel preconditioner for block tridiagonal matrices», in: Parallel Computing: Trends and Applications (G.R. Joubert, D. Trystram, F.J. Peters and D.J. Evans eds), Elsevier Science Publishers B.V., Amsterdam, pp. 113-120, 1994.

15. E. Galligani, V. Ruggiero, «Analysis of splitting methods for solving block tridiagonal linear systems», in: Proceedings of 2nd International Conference on Software for Multiprocessors and Supercomputers, Theory, Practice, Experience (V.P. Ivannikov, V.A. Serebriakov eds), Russian Academy of Sciences, Moscow, pp. 406-416, 1994.

16. E. Galligani, V. Ruggiero, «Implementation of splitting methods for solving block tridiagonal linear systems on transputers», in: Proceedings of 3rd EuroMicro Workshop on Parallel and Distributed Processing (M. Valero, A. Gonzalez eds), IEEE Compute Society Press, Los Alamitos, pp. 409-415, 1995.

17. E. Galligani, V. Ruggiero, «A polynomial preconditioner for block tridiagonal matrices», Taylor & Francis, Parallel Algorithms and Applications, vol. 3, pp. 227-237, 1994.

18. E. Galligani, V. Ruggiero, «The two-stage arithmetic mean method», Elsevier, Applied Mathematics and Computation, vol. 85, pp. 245-264, 1997.

19. K. Hwang, F.A. Briggs, Computer Architecture and Parallel Processing, McGraw-Hill, New York, 1984.

20. C.T. Kelley, Iterative Methods for Linear and Nonlinear Equations, SIAM, Philadelphia, 1995.

21. N.K. Madsen, G.H. Rodrigue, «Odd-even reduction for pentadiagonal matrices», in: Parallel Computers-Parallel Mathematics, Proceedings of IMACS-GI (M. Feilmeier ed.), North-Holland, Amsterdam, pp. 103-106, 1977.

22. G.H. Meyer, «The numerical solution of quasilinear elliptic equations», in: Numerical Solution of Systems of Nonlinear Algebraic Equations (G. Byrne, C.A. Hall eds.), Academic Press, New York, pp. 27-61, 1973.

23. J.M. Ortega, Introduction to Parallel and Vector Solution of Linear Systems, Plenum Press, New York, 1988.

24. J.M. Ortega, Numerical Analysis: A Second Course, SIAM, Philadelphia, 1990.

25. J.M. Ortega, W.C. Rheinboldt, Iterative Solution of Nonlinear Equations in Several Variables, SIAM, Philadelphia, 2000.

26. V. Ruggiero, E. Galligani, «An iterative method for large sparse linear systems on a vector computer», Elsevier, Computers Mathematics with Applications, vol. 20, no. 1, pp. 25-28, 1990.

27. V. Ruggiero, E. Galligani, «A parallel algorithm for solving block tridiagonal linear systems», Elsevier, Computers Mathematics with Applications, vol. 24, no. 4, pp. 15-21, 1992.

28. J. Thomas, Numerical Partial Differential Equations: Finite Difference Methods, Springer, New York, 1995.

29. H.A. van der Vorst, «Bi-CGSTAB: a fast and smoothly converging variant of Bi-CG for the solution of nonsymmetric linear systems», SIAM Journal on Scientific and Statistical Computing, vol. 13, pp. 631-644, 1992.

30. R.S. Varga, Matrix Iterative Analysis, Second Edition, Springer, Berlin, 2000.

Chapter 9

STOCHASTIC MODELING OF REPAIRABLE REDUNDANT SYSTEM COMPRISING ONE BIG UNIT AND THREE SMALL DISSIMILAR UNITS

Ibrahim Yusuf, Nafiu Hussaini

Department of Mathematical Sciences, Faculty of Science, Bayero University, Kano, Nigeria

ABSTRACT

This paper deals with the stochastic modeling of system comprising two subsystems A and B in series. Subsystem A consists three active parallel units. Failure time and repair time are assumed exponential. We developed explicit expressions for mean time to system failure (MTSF), system availability, busy period and profit function using Kolmogorov's forward equations method and perform graphical analysis to see the behavior of failure rates and repair rates on measures of system effectiveness such MTSF, system availability and profit function.

INTRODUCTION

Stochastic models of redundant systems as well as methods of evaluating system reliability indices such as mean time to system failure (MTSF), system availability, busy period of repairman, profit analysis, etc have been researched in order to improve the system effectiveness.

There are systems of three units in which two units are sufficient to perform the entire function of the system. Such systems are called 2-out-of-3 redundant systems. These systems have wide application in the real world. The communication system with three transmitters can be sited as a good example of 2-out-of-3 redundant system. Many research results have been reported on reliability of 2-out-of-3 redundant systems. For example, Chander and Bhardwaj[1], analyzed reliability models for 2-out-of-3 redundant system subject to conditional arrival time of the server. Chander and Bhardwaj[2] present reliability and economic analysis of 2-out-of-3 redundant system with priority to repair. Bhardwaj and Malik[3] studied MTSF and cost effectiveness of 2-out-of-3 cold standby system with probability of repair and inspection. Taneja el al[4] deals with the reliability and cost benefit analysis of a system consisting o a big unit and two identical small units. A single repair facility appears and disappears from the system randomly with constant rates, Malik et al[5] analyzed two reliability models for a system of non identical units original and duplicate using regenerative point technique., Mahmoud and Moshrefa[6] deal with the study of the stochastic analysis of a two unit cold standby system considering hardware failure, human error failure and preventive maintenance, Yusuf and Bala[7], studied stochastic two models of two unit parallel system. In model I, the system can be normal, deterioration (slow, mild or fast deterioration), failure whereas in model II, the system can either be in normal of failure modes. Using linear first order linear differential equations, various measures of system effectiveness such as mean time to system failure (MTSF) and availability are obtained to see the effect of deterioration on such measures, Kumar and Kadyan[8] deal with profit analysis of two unit non identical system with degradation and replacement while Sureria et al[9] studied cost benefit analysis of a computer system with priority to software replacement over hardware repair, Bhardwaj and Malik[15] developed two models for 2-out-of-3 system to study cost benefit analysis using semi-Markov and regenerative process.

Objective

In this paper, we study a system comprising of two subsystems A and B in series. Subsystem A consists of three active parallel units while subsystem B is a single unit. The system is attended by four repairmen and considered up when: (1) all the units of subsystem A and subsystem B are working (2) two units of subsystem A and subsystem B are working. The system is down when two units of subsystem A failed or at the failure of subsystem B. We analyzed the system behavior using kolmogorov's forward equation methods. Explicit expression for measures of system effectiveness like mean time to system failure (MTSF), system availability, busy period of repairman, and profit analysis have been developed. The objective is to study the effect of failure and repair rates parameters with respect to subsystems A and B on reliability indices such as MTSF, availability and profit. Graphs were plotted to see the behavior of failure and repair rates on system performance.

Notations

A_{io} Unit i in subsystem A is operational $i = 1, 2, 3$

A_{Ri} Failed unit in subsystem A under type i repair

A_{iG} Unit i in subsystem A is good

B_o Subsystem B is operational

B_{R4} Subsystem B is failed and under type 4 repair

β_i Type i failure rate of unit A_i in subsystem A

α_i Type i repair rate of unit A_i in subsystem A

λ Failure rate of subsystem B

μ Repair rate of subsystem B

Model Description and Assumptions

1. The system consist of two non identical subsystems A and B

2. Subsystem A consist three active parallel units

3. Units in subsystem A and subsystem B can have two modes: operation and failure

4. The system is attended by four repairmen

5. The system is down when two units of subsystem A failed or at the failure of subsystem B

6. The system is up when all the units of subsystem A and subsystem B are operational or two units of subsystem A and subsystem B are operational

7. Units in subsystem A suffer three types of failures while subsystem B suffer one type of failure

8. Failure rates and repair rates are constant

State of the System

Up states:

$$S_0(A_{10}, A_{20}, A_{30}, B_O)$$
$$S_1(A_{R1}, A_{20}, A_{30}, B_O)$$
$$S_2(A_{10}, A_{R2}, A_{30}, B_O)$$
$$S_3(A_{10}, A_{20}, A_{R3}, B_O)$$

Failed states:

$$S_4(A_{1O}, A_{2O}, A_{3O}, B_{R4})$$
$$S_5(A_{R1}, A_{R2}, A_{3G}, B_G)$$
$$S_6(A_{R1}, A_{2G}, A_{3G}, B_{R4}),$$
$$S_7(A_{1G}, A_{R2}, A_{R3}, B_G)$$
$$S_8(A_{1G}, A_{R2}, A_{3G}, B_{R4})$$
$$S_9(A_{R1}, A_{2G}, A_{R3}, B_G)$$
$$S_{10}(A_{1G}, A_{2G}, A_{R3}, B_{R4})$$

MODEL FORMULATION

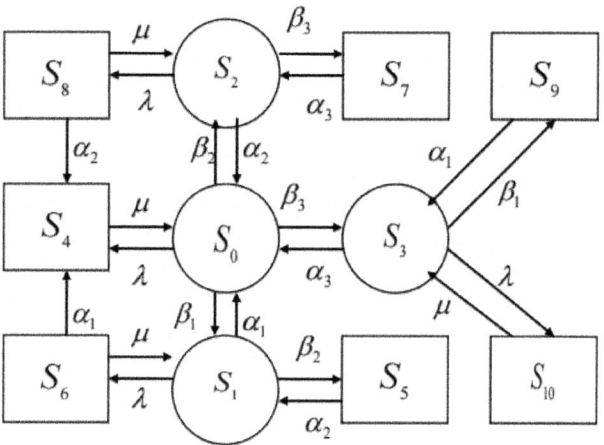

Figure 1: Schematic diagram of the System

Mean Time to System Failure for System

Let P (t) be the probability row vector at time t, then the initial conditions for this problem are as follows:

$$P(0) = [P_0(0), P_1(0), P_2(0), P_3(0), P_4(0), P_5(0), P_6(0), P_7(0), P_8(0), P_9(0), P_{10}(0)] = [1,0,0,0,0,0,0,0,0,0,0]$$

we obtain the following system of differential equations from Figure 1 above:

$$\frac{dP_0(t)}{dt} = -(\lambda + \beta_1 + \beta_2 + \beta_3)P_0(t) + \alpha_1 P_1(t) + \alpha_2 P_2(t) + \alpha_3 P_3(t) + \mu P_4(t)$$

$$\frac{dP_1(t)}{dt} = -(\lambda + \alpha_1 + \beta_2)P_1(t) + \beta_1 P_0(t) + \alpha_2 P_5(t) + \mu P_6(t)$$

$$\frac{dP_2(t)}{dt} = -(\lambda + \alpha_2 + \beta_3)P_2(t) + \beta_2 P_0(t) + \alpha_3 P_7(t) + \mu P_8(t)$$

$$\frac{dP_3(t)}{dt} = -(\lambda + \alpha_3 + \beta_1)P_3(t) + \beta_3 P_0(t) + \alpha_1 P_9(t) + \mu P_{10}(t)$$

$$\frac{dP_4(t)}{dt} = -\mu P_4(t) + \lambda P_0(t) + \alpha_1 P_6(t) + \alpha_2 P_8(t)$$

$$\frac{dP_5(t)}{dt} = -\alpha_2 P_5(t) + \beta_2 P_1(t)$$

$$\frac{dP_6(t)}{dt} = -(\mu + \alpha_1)P_6(t) + \lambda P_1(t)$$

$$\frac{dP_7(t)}{dt} = -\alpha_3 P_7(t) + \beta_2 P_2(t)$$

$$\frac{dP_8(t)}{dt} = -(\mu + \alpha_2)P_8(t) + \lambda P_2(t)$$

$$\frac{dP_9(t)}{dt} = -\alpha_1 P_9(t) + \beta_1 P_3(t)$$

$$\frac{dP_{10}(t)}{dt} = -\mu P_{10}(t) + \lambda P_3(t) \qquad . \tag{1}$$

The differential equations above can be put in matrix form as P=AP where

$$A = \begin{bmatrix}
-(\lambda+\beta_1+\beta_2+\beta_3) & \alpha_1 & \alpha_2 & \alpha_3 & \mu & 0 & 0 & 0 & 0 & 0 & 0 \\
\beta_1 & -(\lambda+\alpha_1+\beta_2) & 0 & 0 & 0 & \alpha_2 & \mu & 0 & 0 & 0 & 0 \\
\beta_1 & 0 & -(\lambda+\alpha_2+\beta_3) & 0 & 0 & 0 & 0 & \alpha_3 & \mu & 0 & 0 \\
\beta_3 & 0 & 0 & -(\lambda+\alpha_3+\beta_1) & 0 & 0 & 0 & 0 & 0 & \alpha_1 & \mu \\
\lambda & 0 & 0 & 0 & -\mu & 0 & \alpha_1 & 0 & \alpha_2 & 0 & 0 \\
0 & \beta_2 & 0 & 0 & 0 & -\alpha_2 & 0 & 0 & 0 & 0 & 0 \\
0 & \lambda & 0 & 0 & 0 & 0 & -(\mu+\alpha_1) & 0 & 0 & 0 & 0 \\
0 & 0 & \beta_3 & 0 & 0 & 0 & 0 & -\alpha_3 & 0 & 0 & 0 \\
0 & 0 & \lambda & 0 & 0 & 0 & 0 & 0 & -(\mu+\alpha_2) & 0 & 0 \\
0 & 0 & 0 & \beta_1 & 0 & 0 & 0 & 0 & 0 & -\alpha_1 & 0 \\
0 & 0 & 0 & \lambda & 0 & 0 & 0 & 0 & 0 & 0 & -\mu
\end{bmatrix}$$

It is difficult to evaluate the transient solutions hence following El-Said[10], Haggag[11], El-Said and Shrbeny [12], and Wang et al[14], we delete the rows and columns of absorbing state of matrix A and take the transpose to produce a new matrix, say Q.

The expected time to reach an absorbing state is obtained from

$$E\left[T_{P(0)\to P(absorbing)}\right] = P(0)(-Q^{-1})\begin{bmatrix}1\\1\\1\\1\\1\end{bmatrix}$$

Where $Q = \begin{bmatrix} -(\lambda+\beta_1+\beta_2+\beta_3) & \beta_1 & \beta_2 & \beta_3 \\ \alpha_1 & -(\lambda+\alpha_1+\beta_2) & 0 & 0 \\ \alpha_2 & 0 & -(\lambda+\alpha_2+\beta_3) & 0 \\ \alpha_3 & 0 & 0 & -(\lambda+\alpha_3+\beta_1) \end{bmatrix}$

This method is successful of the following relations:

$$E\left[T_{P(0)\to P(absorbing)}\right] = P(0)\int_0^\infty e^{At}dt \quad \int_0^\infty e^{At}dt = -A^{-1}, \text{ for } A^{-1}<0$$

Expression for MTSF can therefore be obtain from

$$E\left[T_{P(0)\to P(absorbing)}\right] = MTSF = \frac{N_1}{D_1}$$

$$(2)$$

Where

$N_1 = (\lambda+\alpha_1+\beta_2)(\lambda+\alpha_2+\beta_3)(\lambda+\alpha_3+\beta_1) + \beta_1(\lambda+\alpha_2+\beta_3)(\lambda+\alpha_3+\beta_1) + \beta_1(\lambda+\alpha_1+\beta_2)(\lambda+\alpha_3+\beta_1)$
$\beta_3(\lambda+\alpha_1+\beta_2)(\lambda+\alpha_2+\beta_3)$

$D_1 = \alpha_1\beta_3^2\lambda + \beta_1\beta_2^2\lambda + \alpha_1\beta_1\beta_3^2 + \alpha_2\beta_1\lambda^2 + \alpha_3\beta_2^2\lambda + \alpha_2\beta_2^2\lambda + \beta_1\beta_2\beta_3^2 + \alpha_1\alpha_2\beta_2\lambda + \alpha_1\alpha_2\alpha_3\beta_2 + \alpha_1\alpha_2\beta_1\beta_2 +$
$\alpha_1\alpha_2\alpha_3\lambda + \alpha_1\alpha_3\beta_3\lambda + 2\alpha_1\beta_1\beta_3\lambda + 2\alpha_2\beta_2\beta_3\lambda + 2\alpha_2\beta_1\beta_2\lambda + 2\alpha_3\beta_2\beta_3\lambda + 4\beta_1\beta_2\beta_3\lambda + \alpha_3\beta_2\beta_3\lambda - \alpha_1\alpha_2\alpha_3\beta_1 +$
$\alpha_3\beta_1\beta_2\lambda + \alpha_3\beta_1\beta_2\beta_3 + \alpha_1\alpha_3\beta_2\lambda + \alpha_1\beta_1\beta_2\lambda + \alpha_2\alpha_3\lambda^2 + \alpha_1\alpha_2\lambda^2 + \lambda^4 + \alpha_3\lambda^3 + 2\beta_1\lambda^3 + \alpha_2\lambda^3 + 2\beta_3\lambda^3 + \alpha_1\lambda^3 +$
$2\beta_2\lambda^3 + \beta_1^2\lambda^2 + \beta_2^2\lambda^2 + \beta_3^2\lambda^2 + 2\alpha_1\beta_3\lambda^2 + 2\alpha_2\beta_2\lambda^2 + 3\beta_2\beta_3\lambda^2 + \beta_1^2\beta_3\lambda - \alpha_1\alpha_2\beta_1^2 + \beta_1^2\beta_2\lambda + \beta_1^2\beta_2\beta_3 +$
$\alpha_1\beta_2\lambda^2 + \alpha_2\beta_3\lambda^2 + \alpha_3\beta_1\lambda^2 + 3\beta_1\beta_3\lambda^2 + \alpha_1\alpha_3\lambda^2 + \alpha_1\beta_1\lambda^2 + 2\alpha_3\beta_2\lambda^2 + 3\beta_1\beta_2\lambda^2 + \alpha_3\beta_1\lambda^2 + \alpha_2\alpha_3\beta_2^2 + \alpha_2\beta_1\beta_2^2 +$
$+\beta_2^2\beta_3\lambda + \alpha_3\beta_2^2\beta_3 + \beta_1\beta_2^2\beta_3 + \beta_1\beta_3^2\lambda + \beta_2\beta_3^2\lambda + \alpha_1\beta_2\beta_3\lambda + \alpha_1\alpha_3\beta_2\beta_3 + \alpha_1\beta_1\beta_2\beta_3 + \alpha_2\beta_1\beta_3\lambda + \alpha_1\alpha_2\beta_3\lambda +$
$\alpha_1\alpha_2\beta_1\beta_3 + \alpha_2\beta_2\beta_3\lambda + \alpha_2\beta_1\beta_2\beta_3$

Steady state availability Analysis for System

For the availability case of Figure 1 following El-Said[10], Haggag[11], El-Said and Shrbeny[12], and Wang et al[14], the initial conditions for this system are:

$P(0) = [P_0(0), P_1(0), P_2(0), P_3(0), P_4(0), P_5(0), P_6(0), P_7(0), P_8(0), P_9(0), P_{10}(0)] = [1,0,0,0,0,0,0,0,0,0,0]$

The system of differential equations in for System 1 above can be expressed as:

$$
\begin{bmatrix}
\dot{P_0}(t) \\
\dot{P_1}(t) \\
\dot{P_2}(t) \\
\dot{P_3}(t) \\
\dot{P_4}(t) \\
\dot{P_5}(t) \\
\dot{P_6}(t) \\
\dot{P_7}(t) \\
\dot{P_8}(t) \\
\dot{P_9}() \\
\dot{P_{10}}(t)
\end{bmatrix} =
$$

$$
\begin{bmatrix}
-(\lambda+\beta_1+\beta_2+\beta_3) & \alpha_1 & \alpha_2 & \alpha_3 & \mu & 0 & 0 & 0 & 0 & 0 & 0 \\
\beta_1 & -(\lambda+\alpha_1+\beta_2) & 0 & 0 & 0 & \alpha_2 & \mu & 0 & 0 & 0 & 0 \\
\beta_1 & 0 & -(\lambda+\alpha_2+\beta_3) & 0 & 0 & 0 & 0 & \alpha_3 & \mu & 0 & 0 \\
\beta_3 & 0 & 0 & -(\lambda+\alpha_3+\beta_1) & 0 & 0 & 0 & 0 & 0 & \alpha_1 & \mu \\
\lambda & 0 & 0 & 0 & -\mu & 0 & \alpha_1 & 0 & \alpha_2 & 0 & 0 \\
0 & \beta_2 & 0 & 0 & 0 & -\alpha_2 & 0 & 0 & 0 & 0 & 0 \\
0 & \lambda & 0 & 0 & 0 & 0 & -(\mu+\alpha_1) & 0 & 0 & 0 & 0 \\
0 & 0 & \beta_3 & 0 & 0 & 0 & 0 & -\alpha_3 & 0 & 0 & 0 \\
0 & 0 & \lambda & 0 & 0 & 0 & 0 & 0 & -(\mu+\alpha_2) & 0 & 0 \\
0 & 0 & 0 & \beta_1 & 0 & 0 & 0 & 0 & 0 & -\alpha_1 & 0 \\
0 & 0 & 0 & \lambda & 0 & 0 & 0 & 0 & 0 & 0 & -\mu
\end{bmatrix}
\begin{bmatrix}
P_0(t) \\
P_1(t) \\
P_2(t) \\
P_3(t) \\
P_4(t) \\
P_5(t) \\
P_6(t) \\
P_7(t) \\
P_8(t) \\
P_9(t) \\
P_{10}(t)
\end{bmatrix}
$$

The steady-state availability is given by

$$
A_V(\infty) = P_0(\infty) + P_1(\infty) + P_2(\infty) + P_4(\infty) + P_7(\infty)
\tag{3}
$$

In the steady state, the derivatives of the state probabilities become zero so that

$$
AP = 0
\tag{4}
$$

which in matrix form

$$
\begin{bmatrix}
-(\lambda+\beta_1+\beta_2+\beta_3) & \alpha_1 & \alpha_2 & \alpha_3 & \mu & 0 & 0 & 0 & 0 & 0 & 0 \\
\beta_1 & -(\lambda+\alpha_1+\beta_2) & 0 & 0 & 0 & \alpha_2 & \mu & 0 & 0 & 0 & 0 \\
\beta_1 & 0 & -(\lambda+\alpha_2+\beta_3) & 0 & 0 & 0 & 0 & \alpha_3 & \mu & 0 & 0 \\
\beta_3 & 0 & 0 & -(\lambda+\alpha_3+\beta_1) & 0 & 0 & 0 & 0 & 0 & \alpha_1 & \mu \\
\lambda & 0 & 0 & 0 & -\mu & 0 & \alpha_1 & 0 & \alpha_2 & 0 & 0 \\
0 & \beta_2 & 0 & 0 & 0 & -\alpha_2 & 0 & 0 & 0 & 0 & 0 \\
0 & \lambda & 0 & 0 & 0 & 0 & -(\mu+\alpha_1) & 0 & 0 & 0 & 0 \\
0 & 0 & \beta_3 & 0 & 0 & 0 & 0 & -\alpha_3 & 0 & 0 & 0 \\
0 & 0 & \lambda & 0 & 0 & 0 & 0 & 0 & -(\mu+\alpha_2) & 0 & 0 \\
0 & 0 & 0 & \beta_1 & 0 & 0 & 0 & 0 & 0 & -\alpha_1 & 0 \\
0 & 0 & 0 & \lambda & 0 & 0 & 0 & 0 & 0 & 0 & -\mu
\end{bmatrix}
\begin{bmatrix}
P_0(t) \\
P_1(t) \\
P_2(t) \\
P_3(t) \\
P_4(t) \\
P_5(t) \\
P_6(t) \\
P_7(t) \\
P_8(t) \\
P_{10}(t)
\end{bmatrix} =
\begin{bmatrix}
0 \\
0 \\
0 \\
0 \\
0 \\
0 \\
0 \\
0 \\
0 \\
0 \\
0
\end{bmatrix}
$$

Using the following normalizing condition

$$P_0(\infty) + P_1(\infty) + P_2(\infty) + P_3(\infty) + P_4(\infty) + P_5(\infty) + P_6(\infty) + P_7(\infty) + P_8(\infty) + P_9(\infty) + P_{10}(\infty) = 1$$

(5)

We substitute (5) in any of the redundant rows in (4) to give

$$
\begin{bmatrix}
-(\lambda+\beta_1+\beta_2+\beta_3) & \alpha_1 & \alpha_2 & \alpha_3 & \mu & 0 & 0 & 0 & 0 & 0 & 0 \\
\beta_1 & -(\lambda+\alpha_1+\beta_2) & 0 & 0 & 0 & \alpha_2 & \mu & 0 & 0 & 0 & 0 \\
\beta_1 & 0 & -(\lambda+\alpha_2+\beta_3) & 0 & 0 & 0 & 0 & \alpha_3 & \mu & 0 & 0 \\
\beta_3 & 0 & 0 & -(\lambda+\alpha_3+\beta_1) & 0 & 0 & 0 & 0 & 0 & \alpha_1 & \mu \\
\lambda & 0 & 0 & 0 & -\mu & 0 & \alpha_1 & 0 & \alpha_2 & 0 & 0 \\
0 & \beta_2 & 0 & 0 & 0 & -\alpha_2 & 0 & 0 & 0 & 0 & 0 \\
0 & \lambda & 0 & 0 & 0 & 0 & -(\mu+\alpha_1) & 0 & 0 & 0 & 0 \\
0 & 0 & \beta_3 & 0 & 0 & 0 & 0 & -\alpha_3 & 0 & 0 & 0 \\
0 & 0 & \lambda & 0 & 0 & 0 & 0 & 0 & -(\mu+\alpha_2) & 0 & 0 \\
0 & 0 & 0 & \beta_1 & 0 & 0 & 0 & 0 & 0 & -\alpha_1 & 0 \\
1 & 1 & 1 & 1 & 1 & 1 & 1 & 1 & 1 & 1 & 1
\end{bmatrix}
\begin{bmatrix}
P_0(\infty) \\ P_1(\infty) \\ P_2(\infty) \\ P_3(\infty) \\ P_4(\infty) \\ P_5(\infty) \\ P_6(\infty) \\ P_7(\infty) \\ P_8(\infty) \\ P_9(\infty) \\ P_{10}(\infty)
\end{bmatrix}
=
\begin{bmatrix}
0 \\ 0 \\ 0 \\ 0 \\ 0 \\ 0 \\ 0 \\ 0 \\ 0 \\ 0 \\ 1
\end{bmatrix}
$$

We solve for the system of equations in the matrix above to obtain the steady-state probabilities

$$P_0(\infty), P_1(\infty), P_2(\infty), P_3(\infty)$$

$$A_V = \frac{N_2}{D_2}$$

Where

$$N_2 =$$

$$\alpha_1\alpha_2\alpha_3\mu(\mu^2 + 2\mu\lambda + \alpha_1\mu + \alpha_2\mu + \lambda^2 + \alpha_1\lambda + \alpha_2\lambda + \alpha_1\alpha_2) + \alpha_2\alpha_3\beta_1\mu(\mu^2 + \mu\lambda + \alpha_1\mu + \alpha_2\mu + \alpha_1\lambda + \alpha_1\alpha_2) +$$

$$\alpha_1\alpha_3\beta_1\mu(\mu^2 + \mu\lambda + \alpha_1\mu + \alpha_2\mu + \alpha_2\lambda + \alpha_1\alpha_2) - \alpha_1\alpha_2\mu(-\beta_2\mu + \beta_1\mu^2 - \beta_3\mu^2 + 2\beta_1\mu\lambda - 2\beta_2\mu\lambda - 2\beta_3\mu\lambda - \alpha_1\beta_2\mu +$$
$$\alpha_1\beta_1\mu - \alpha_1\beta_3\mu - \alpha_2\beta_2\mu + \alpha_2\beta_1\mu - \alpha_2\beta_3\mu - \beta_2\lambda^2 + \beta_1\lambda^2 - \beta_3\lambda^2 - \alpha_1\beta_2\lambda + \alpha_1\beta_1\lambda - \alpha_1\beta_3\lambda - \alpha_2\beta_2\lambda + \alpha_2\beta_1\lambda -$$
$$\alpha_2\beta_3\lambda - \alpha_1\alpha_2\beta_2 + \alpha_1\alpha_2\beta_1 - \alpha_1\alpha_2\beta_3)$$

$$D_2 =$$

$$\alpha_1^2\alpha_2^2\alpha_3\lambda + 2\alpha_1\alpha_3\beta_1\mu^2\lambda + \alpha_1\alpha_3\beta_1\mu\lambda^2 - 2\alpha_2\beta_1^2\mu^2\lambda - \alpha_2\beta_1^2\mu\lambda^2 - \alpha_1\alpha_2\beta_1^2\mu\lambda + \alpha_1^2\alpha_3\beta_1\mu\lambda + \alpha_3\beta_1\beta_2\mu^2\lambda +$$
$$\alpha_1\alpha_3\beta_1\beta_2\mu\lambda + \alpha_1^2\alpha_2^2\alpha_3\beta_2 + \alpha_1\alpha_2^2\alpha_3\beta_2\lambda - \alpha_1^2\alpha_2^2\alpha_3\beta_1 + \alpha_1\alpha_2\alpha_3\mu^3 + \alpha_1^2\alpha_2\alpha_3\mu^2 + \alpha_1\alpha_2^2\alpha_3\mu^2 + \alpha_1^2\alpha_2^2\alpha_3\mu +$$
$$\alpha_1\alpha_2\beta_1\mu^3 + \alpha_1\alpha_2\beta_2\mu^3 + \alpha_1\alpha_3\beta_1\mu^3 + \alpha_1\alpha_2\alpha_3\beta_1\beta_2\mu + \alpha_2^2\alpha_3\beta_1\mu\lambda + 2\alpha_2\alpha_3\beta_1\mu^2\lambda + 3\alpha_1\alpha_2\beta_1\mu^2\lambda + \alpha_2^2\beta_1\beta_2\mu^2 +$$
$$\alpha_1\alpha_2\beta_1\beta_2\mu^2 + \alpha_1\alpha_2^2\beta_1\mu^2 + 3\alpha_1\alpha_2\beta_2\mu\lambda^2 + 2\alpha_1\alpha_2\alpha_3\beta_2\mu\lambda + 2\alpha_1^2\alpha_2\alpha_3\mu\lambda + \alpha_1^2\alpha_2^2\beta_1\mu + 2\alpha_1^2\alpha_2\beta_1\mu\lambda + \alpha_1^2\alpha_3\beta_1\beta_3\mu -$$
$$3\alpha_1\alpha_2\beta_1\mu\lambda^2 + \alpha_2\beta_1\beta_3\mu\lambda^2 + \alpha_2\beta_1\beta_3\mu\lambda^2 - 2\alpha_1\alpha_2^2\beta_1\mu\lambda + \alpha_2^2\beta_1\beta_2\mu\lambda + \alpha_2^2\beta_1\beta_2\mu\lambda - \alpha_1^2\alpha_2^2\beta_1\mu + \alpha_1^2\alpha_2^2\beta_2\mu +$$
$$\alpha_1^2\alpha_2\alpha_3\beta_2\mu - \alpha_1\alpha_2\beta_1^2\mu + \alpha_1\alpha_2^2\alpha_3\beta_2\mu + \alpha_1\alpha_2^2\beta_1\beta_2\mu + 2\alpha_1\alpha_2^2\alpha_3\mu\lambda + 2\alpha_1\alpha_2\alpha_3\beta_1\mu\lambda + 2\alpha_1\alpha_2\beta_1\beta_3\mu\lambda +$$
$$2\alpha_1\alpha_2^2\beta_1\mu\lambda + 2\alpha_1\alpha_2^2\beta_2\mu\lambda + \alpha_2\alpha_3\beta_1\mu\lambda^2 + 2\alpha_1^2\alpha_2\beta_2\mu\lambda + \alpha_1\alpha_2^2\beta_1\beta_2\mu - \alpha_1^2\alpha_2^2\beta_1\lambda + \alpha_1^2\alpha_2^2\beta_3\lambda + \alpha_1^2\alpha_2^2\beta_2\lambda +$$
$$\alpha_1^2\alpha_2\alpha_3\beta_2\lambda + \alpha_1^2\alpha_2\alpha_3\lambda^2 - \alpha_1^2\alpha_2\beta_1\lambda^2 + \alpha_1^2\alpha_2\beta_3\lambda^2 + \alpha_1^2\alpha_2\beta_2\lambda^2 + \alpha_1\alpha_2^2\alpha_3\lambda - \alpha_1\alpha_2^2\beta_1\lambda^2 + \alpha_1\alpha_2^2\beta_3\lambda^2 +$$
$$\alpha_1\alpha_2^2\beta_2\lambda^2 + \alpha_1\alpha_2\alpha_3\beta_1\lambda^2 + \alpha_1\alpha_2\alpha_3\beta_2\lambda^2 + \alpha_1\alpha_2\alpha_3\lambda^3 - \alpha_1\alpha_2\beta_1\lambda^3 + \alpha_1\alpha_2\beta_3\lambda^3 + \alpha_1\alpha_2\beta_2\lambda^3 + \alpha_1\alpha_3\beta_1\beta_2\mu^2 +$$
$$\alpha_1\alpha_2\alpha_3\beta_1\mu^2 + \alpha_1\beta_1\beta_3\mu^2\lambda + \alpha_1^2\alpha_3\beta_1\mu^2 + \alpha_2^2\beta_1\beta_2\mu^2 + \alpha_1^2\alpha_3\beta_1\beta_2\mu^2 - 3\alpha_1\alpha_2\beta_1\mu^2\lambda + 2\alpha_2\beta_1\beta_2\mu^2\lambda + 2\alpha_2\beta_1\beta_2\mu^2\lambda +$$
$$\alpha_1^2\alpha_2\beta_2\mu^2 - \alpha_1\alpha_2^2\beta_1\mu^2 - \alpha_1\alpha_2\beta_1^2\mu^2 + \alpha_2^2\alpha_3\beta_1\mu^2 - \alpha_2^2\beta_1^2\mu^2 + \alpha_1^2\alpha_2\beta_3\mu^2 + 3\alpha_1\alpha_2\beta_2\mu^2\lambda - \alpha_1\alpha_2\beta_1\mu^2 +$$
$$\alpha_1\alpha_2^2\beta_2\mu^2 + \alpha_1^2\beta_1\beta_3\mu^2 + \alpha_1\alpha_2\beta_1\beta_2\mu\lambda - \alpha_2^2\beta_1^2\mu\lambda - 2\alpha_1^2\alpha_2\beta_1\mu\lambda + 3\alpha_1\alpha_2\beta_1\mu\lambda^2 + 3\alpha_1\alpha_2\alpha_3\mu\lambda^2 + \alpha_1\beta_1\beta_3\mu^3 +$$
$$\alpha_2\alpha_3\beta_1\mu^3 - \alpha_2\beta_1^2\mu^3 + \alpha_2\beta_1\beta_3\mu^3 + \alpha_2\beta_1\beta_2\mu^3 + \alpha_3\beta_1\beta_2\mu^3 + 2\alpha_1\alpha_2\beta_1\beta_2\mu^2 + \alpha_1\alpha_2\alpha_3\beta_2\mu^2 + 3\alpha_1\alpha_2\alpha_3\mu^2\lambda - \alpha_1\alpha_2\beta_1\mu^3$$

Busy Period Analysis

Using the same initial conditions as for the reliability case:

$$P(0) = [P_0(0), P_1(0), P_2(0), P_3(0), P_4(0), P_5(0), P_6(0), P_7(0), P_8(0), P_9(0), P_{10}(0)]$$
$$= [1, 0, 0, 0, 0, 0, 0, 0, 0, 0, 0]$$

The differential equations can be expressed as

$$
\begin{bmatrix}
\dot{P_0}(t) \\
\dot{P_1}(t) \\
\dot{P_2}(t) \\
\dot{P_3}(t) \\
\dot{P_4}(t) \\
\dot{P_5}(t) \\
\dot{P_6}(t) \\
\dot{P_7}(t) \\
\dot{P_8}(t) \\
\dot{P_9}() \\
\dot{P_{10}}(t)
\end{bmatrix} =
$$

$$
\begin{bmatrix}
-(\lambda+\beta_1+\beta_2+\beta_3) & \alpha_1 & \alpha_2 & \alpha_3 & \mu & 0 & 0 & 0 & 0 & 0 & 0 \\
\beta_1 & -(\lambda+\alpha_1+\beta_2) & 0 & 0 & 0 & \alpha_2 & \mu & 0 & 0 & 0 & 0 \\
\beta_1 & 0 & -(\lambda+\alpha_2+\beta_3) & 0 & 0 & 0 & 0 & \alpha_3 & \mu & 0 & 0 \\
\beta_3 & 0 & 0 & -(\lambda+\alpha_3+\beta_1) & 0 & 0 & 0 & 0 & 0 & \alpha_1 & \mu \\
\lambda & 0 & 0 & 0 & -\mu & 0 & \alpha_1 & 0 & \alpha_2 & 0 & 0 \\
0 & \beta_2 & 0 & 0 & 0 & -\alpha_2 & 0 & 0 & 0 & 0 & 0 \\
0 & \lambda & 0 & 0 & 0 & 0 & -(\mu+\alpha_1) & 0 & 0 & 0 & 0 \\
0 & 0 & \beta_3 & 0 & 0 & 0 & 0 & -\alpha_3 & 0 & 0 & 0 \\
0 & 0 & \lambda & 0 & 0 & 0 & 0 & 0 & -(\mu+\alpha_2) & 0 & 0 \\
0 & 0 & 0 & \beta_1 & 0 & 0 & 0 & 0 & 0 & -\alpha_1 & 0 \\
0 & 0 & 0 & \lambda & 0 & 0 & 0 & 0 & 0 & 0 & -\mu
\end{bmatrix}
\begin{bmatrix}
P_0(t) \\
P_1(t) \\
P_2(t) \\
P_3(t) \\
P_4(t) \\
P_5(t) \\
P_6(t) \\
P_7(t) \\
P_8(t) \\
P_9(t) \\
P_{10}(t)
\end{bmatrix}
$$

In the steady state, the derivatives of the state probabilities become zero this will enable us to compute steady state busy:

$$B(\infty) = 1 - P_0(\infty) \qquad\qquad (6)$$

$AP = 0$

$$\begin{bmatrix} -(\lambda+\beta_1+\beta_2+\beta_3) & \alpha_1 & \alpha_2 & \alpha_3 & \mu & 0 & 0 & 0 & 0 & 0 & 0 \\ \beta_1 & -(\lambda+\alpha_1+\beta_2) & 0 & 0 & 0 & \alpha_2 & \mu & 0 & 0 & 0 & 0 \\ \beta_1 & 0 & -(\lambda+\alpha_2+\beta_3) & 0 & 0 & 0 & 0 & \alpha_3 & \mu & 0 & 0 \\ \beta_3 & 0 & 0 & -(\lambda+\alpha_3+\beta_1) & 0 & 0 & 0 & 0 & 0 & \alpha_1 & \mu \\ \lambda & 0 & 0 & 0 & -\mu & 0 & \alpha_1 & 0 & \alpha_2 & 0 & 0 \\ 0 & \beta_2 & 0 & 0 & 0 & -\alpha_2 & 0 & 0 & 0 & 0 & 0 \\ 0 & \lambda & 0 & 0 & 0 & 0 & -(\mu+\alpha_1) & 0 & 0 & 0 & 0 \\ 0 & 0 & \beta_3 & 0 & 0 & 0 & 0 & -\alpha_3 & 0 & 0 & 0 \\ 0 & 0 & \lambda & 0 & 0 & 0 & 0 & 0 & -(\mu+\alpha_2) & 0 & 0 \\ 0 & 0 & 0 & \beta_1 & 0 & 0 & 0 & 0 & 0 & -\alpha_1 & 0 \\ 0 & 0 & 0 & \lambda & 0 & 0 & 0 & 0 & 0 & 0 & -\mu \end{bmatrix} \begin{bmatrix} P_0(t) \\ P_1(t) \\ P_2(t) \\ P_3(t) \\ P_4(t) \\ P_5(t) \\ P_6(t) \\ P_7(t) \\ P_8(t) \\ P_9(t) \\ P_{10}(t) \end{bmatrix} = \begin{bmatrix} 0 \\ 0 \\ 0 \\ 0 \\ 0 \\ 0 \\ 0 \\ 0 \\ 0 \\ 0 \\ 0 \end{bmatrix}$$

We solve for $P_0(\infty)$

Using the following normalizing condition

$$P_0(\infty) + P_1(\infty) + P_2(\infty) + P_3(\infty) + P_4(\infty) + P_5(\infty) + P_6(\infty) + P_7(\infty) + P_8(\infty) + P_9(\infty) + P_{10}(\infty) = 1$$

We substitute (5) in any of the redundant rows in (4) to give

$$\begin{bmatrix} -(\lambda+\beta_1+\beta_2+\beta_3) & \alpha_1 & \alpha_2 & \alpha_3 & \mu & 0 & 0 & 0 & 0 & 0 & 0 \\ \beta_1 & -(\lambda+\alpha_1+\beta_2) & 0 & 0 & 0 & \alpha_2 & \mu & 0 & 0 & 0 & 0 \\ \beta_1 & 0 & -(\lambda+\alpha_2+\beta_3) & 0 & 0 & 0 & 0 & \alpha_3 & \mu & 0 & 0 \\ \beta_3 & 0 & 0 & -(\lambda+\alpha_3+\beta_1) & 0 & 0 & 0 & 0 & 0 & \alpha_1 & \mu \\ \lambda & 0 & 0 & 0 & -\mu & 0 & \alpha_1 & 0 & \alpha_2 & 0 & 0 \\ 0 & \beta_2 & 0 & 0 & 0 & -\alpha_2 & 0 & 0 & 0 & 0 & 0 \\ 0 & \lambda & 0 & 0 & 0 & 0 & -(\mu+\alpha_1) & 0 & 0 & 0 & 0 \\ 0 & 0 & \beta_3 & 0 & 0 & 0 & 0 & -\alpha_3 & 0 & 0 & 0 \\ 0 & 0 & \lambda & 0 & 0 & 0 & 0 & 0 & -(\mu+\alpha_2) & 0 & 0 \\ 0 & 0 & 0 & \beta_1 & 0 & 0 & 0 & 0 & 0 & -\alpha_1 & 0 \\ 1 & 1 & 1 & 1 & 1 & 1 & 1 & 1 & 1 & 1 & 1 \end{bmatrix} \begin{bmatrix} P_0(\infty) \\ P_1(\infty) \\ P_2(\infty) \\ P_3(\infty) \\ P_4(\infty) \\ P_5(\infty) \\ P_6(\infty) \\ P_7(\infty) \\ P_8(\infty) \\ P_9(\infty) \\ P_{10}(\infty) \end{bmatrix} = \begin{bmatrix} 0 \\ 0 \\ 0 \\ 0 \\ 0 \\ 0 \\ 0 \\ 0 \\ 0 \\ 0 \\ 1 \end{bmatrix}$$

The steady state busy period $B(\infty)$ is therefore

$$B(\infty) = \frac{N_3}{D_2}$$

$$N_3 =$$

$\alpha_1\alpha_2\beta_3\mu^3 + \alpha_1\alpha_2\beta_2\mu^3 + \alpha_1\alpha_3\beta_1\mu^3 + \alpha_1\alpha_2\alpha_3\beta_1\beta_2\mu + \alpha_1^2\alpha_3\beta_1\lambda + 2\alpha_1\alpha_3\beta_1\mu^2\lambda + 3\alpha_1\alpha_2\beta_3\mu^2\lambda + \alpha_1^2\beta_1\beta_3\mu^2 +$
$\alpha_1\alpha_2\beta_1\beta_2\mu^2 + \alpha_1\alpha_2^2\beta_3\mu^2 + 3\alpha_1\alpha_2\beta_2\mu\lambda^2 + 2\alpha_1\alpha_2\alpha_3\beta_1\mu\lambda + \alpha_1^2\alpha_3\mu\lambda + \alpha_1^2\alpha_2^2\beta_3\mu + 2\alpha_1^2\alpha_2\beta_3\mu\lambda + \alpha_1^2\alpha_2\beta_3\beta_1\mu -$
$3\alpha_1\alpha_2\beta_1\mu\lambda^2 + \alpha_2\beta_1\beta_3\mu\lambda^2 + \alpha_2\beta_1\beta_3\mu\lambda^2 - 2\alpha_1\alpha_2^2\beta_1\mu\lambda + \alpha_2^2\beta_1\beta_3\mu\lambda + \alpha_2^2\beta_1\beta_2\mu\lambda - \alpha_1^2\alpha_2^2\beta_1\mu + \alpha_1^2\alpha_2^2\beta_2\mu +$
$\alpha_1^2\alpha_2\alpha_3\beta_2\mu - \alpha_1\alpha_2^2\beta_1^2\mu + \alpha_1\alpha_2^2\beta_2\mu + \alpha_1\alpha_2^2\beta_1\beta_3\mu + \alpha_1\alpha_2^2\mu\lambda + 2\alpha_1\alpha_2\alpha_3\beta_1\mu\lambda + 2\alpha_1\alpha_2\beta_3\mu\lambda +$
$2\alpha_1\alpha_2^2\beta_3\mu\lambda + 2\alpha_1\alpha_2^2\beta_2\mu\lambda + \alpha_2\alpha_3\beta_1\mu\lambda^2 + 2\alpha_1^2\alpha_2\beta_2\mu\lambda + \alpha_1\alpha_2^2\beta_1\beta_2\mu - \alpha_1^2\alpha_2^2\beta_1\lambda + \alpha_1^2\alpha_2^2\beta_3\lambda + \alpha_1^2\alpha_2^2\beta_2\lambda +$

$\alpha_1^2\alpha_2\alpha_3\beta_2\lambda + \alpha_1^2\alpha_2\alpha_3\lambda^2 - \alpha_1^2\alpha_2\beta_1\lambda^2 + \alpha_1^2\alpha_2\beta_3\lambda^2 + \alpha_1^2\alpha_2\beta_2\lambda^2 + \alpha_1\alpha_2^2\alpha_3\lambda^2 - \alpha_1\alpha_2^2\beta_1\lambda^2 + \alpha_1\alpha_2^2\beta_3\lambda^2 +$
$\alpha_1\alpha_2^2\beta_2\lambda^2 + \alpha_1\alpha_2\alpha_3\beta_1\lambda^2 + \alpha_1\alpha_2\alpha_3\beta_2\lambda^2 + \alpha_1\alpha_2\alpha_3\lambda^3 - \alpha_1\alpha_2\beta_1\lambda^3 + \alpha_1\alpha_2\beta_3\lambda^3 + \alpha_1\alpha_2\beta_2\lambda^3 + \alpha_1\alpha_3\beta_1\beta_2\mu^2 +$
$\alpha_1\alpha_2\alpha_3\beta_1\mu^2 + \alpha_1\beta_1\beta_3\mu^2\lambda + \alpha_1^2\alpha_3\beta_1\mu^2 + \alpha_2^2\beta_1\beta_2\mu^2 + \alpha_1^2\alpha_3\beta_1\beta_2\mu^2 - 3\alpha_1\alpha_2\beta_1\mu^2\lambda + 2\alpha_2\beta_1\beta_3\mu^2\lambda + 2\alpha_1\beta_1\beta_2\mu^2\lambda +$
$\alpha_1^2\alpha_2\beta_2\mu^2 - \alpha_1\alpha_2^2\beta_1\mu^2 - \alpha_1\alpha_2\beta_1^2\mu^2 + \alpha_1^2\alpha_3\beta_1\mu^2 - \alpha_2^2\beta_1^2\mu^2 + \alpha_1^2\alpha_2\beta_3\mu^2 + 3\alpha_1\alpha_2\beta_2\mu^2\lambda - \alpha_1^2\alpha_2\beta_1\mu^2 +$
$\alpha_1\alpha_2^2\beta_2\mu^2 + \alpha_1^2\beta_1\beta_3\mu^2 + \alpha_1\alpha_2\beta_1\beta_2\mu\lambda - \alpha_2^2\beta_1^2\mu\lambda - 2\alpha_1^2\alpha_2\beta_1\mu\lambda + 3\alpha_1\alpha_2\beta_3\mu\lambda^2 + 2\alpha_1\alpha_2\alpha_3\mu\lambda^2 + \alpha_1\beta_1\beta_2\mu\lambda^3 +$
$\alpha_2\alpha_3\beta_1\mu\lambda^3 - \alpha_2\beta_1^2\mu^3 + \alpha_2\beta_1\beta_3\mu^3 + \alpha_2\beta_1\beta_2\mu^3 + \alpha_3\beta_1\beta_2\mu^3 + 2\alpha_1\alpha_2\beta_1\beta_3\mu^2 + \alpha_1\alpha_2\alpha_3\beta_2\mu^2 + \alpha_1\alpha_2\alpha_3\mu^2\lambda - \alpha_1\alpha_2\beta_1\mu^3 +$
$\alpha_1^2\alpha_2\alpha_3\lambda + 2\alpha_1\alpha_3\beta_1\mu^2\lambda + \alpha_1\alpha_3\beta_1\mu\lambda^2 - 2\alpha_2\beta_1^2\mu^2\lambda - \alpha_2\beta_1^2\mu\lambda^2 - \alpha_1\alpha_2\beta_1^2\mu\lambda + \alpha_1^2\alpha_3\beta_1\mu\lambda + \alpha_3\beta_1\beta_2\mu^2\lambda +$
$$\alpha_1\alpha_3\beta_1\beta_2\mu\lambda + \alpha_1^2\alpha_2^2\alpha_3\beta_2 + \alpha_1\alpha_2^2\alpha_3\beta_2\lambda - \alpha_1^2\alpha_2^2\alpha_3\beta_1$$

Profit Analysis

Following El-Said[10], Haggag[11], El-said and sherbeny[13] and Wang et al[14], the expected profit per unit time incurred to the system in the steady-state is given by:

Profit =total revenue generated from system using - total cost due to repair of failed unit or subsystem B

$$PF = C_0 A_{V2}(\infty) - C_1 B_2(\infty)$$

(6)

Where PF: is the profit incurred to the system

C_0: is the revenue per unit up time of the system

C_1: is the cost per unit time which the system is under repair

RESULTS

The following particular cases are considered:

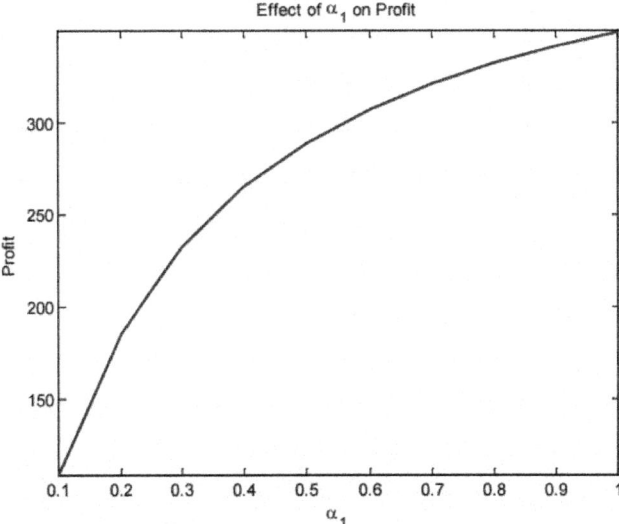

Figure 2: effect of α_1 on Profit

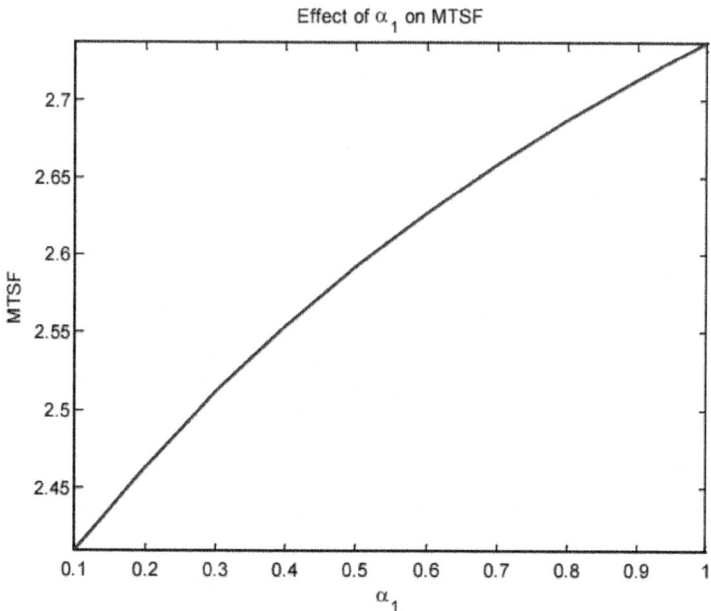

Figure 3: effect of α_1 on MTSF

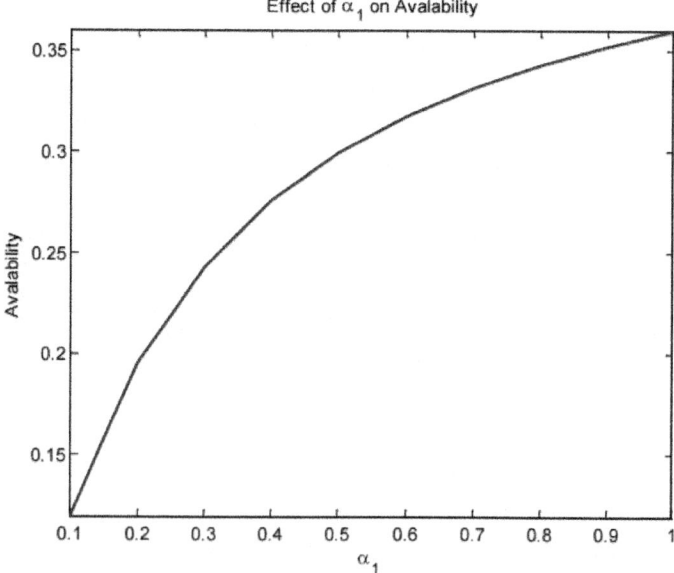

Figure 4: effect of α_1 on system availability

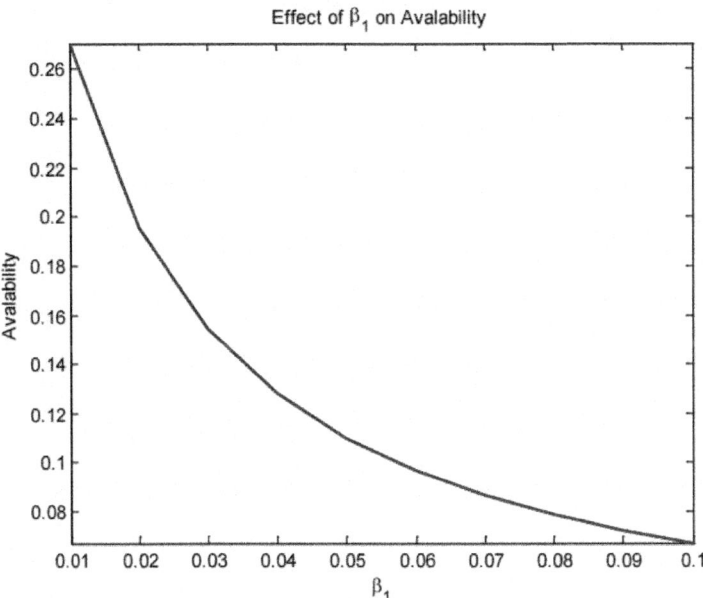

Figure 5: effect of β_1 on system availability

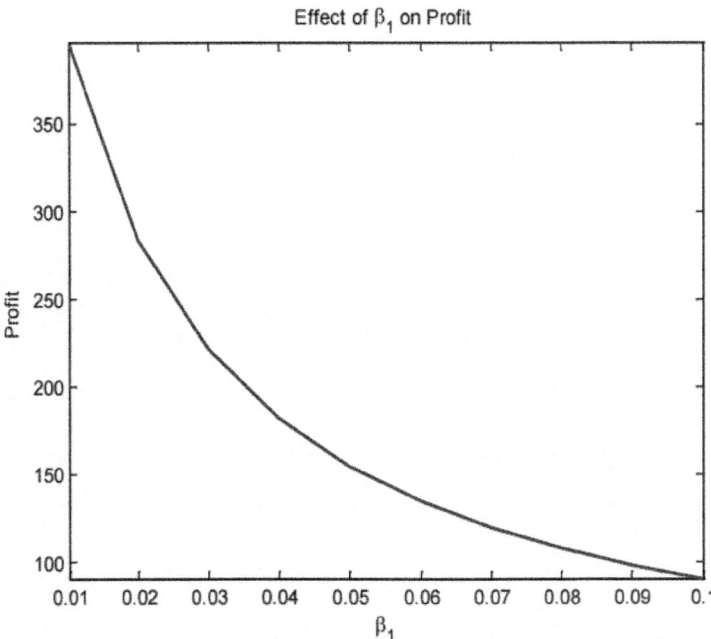

Figure 6: effect of β_1 on Profit

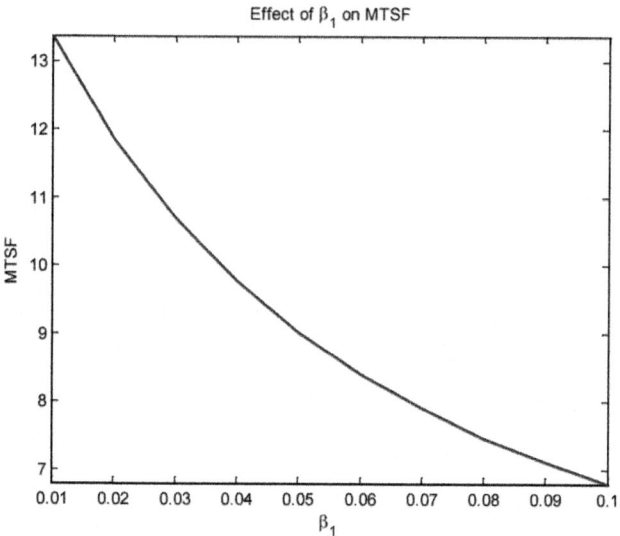

Figure 7: effect of β1 on MTSF

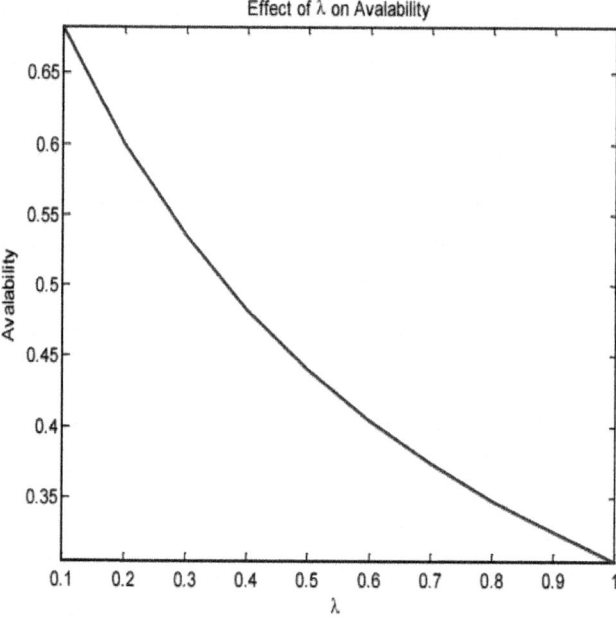

Figure 8: effect of λ on system availability

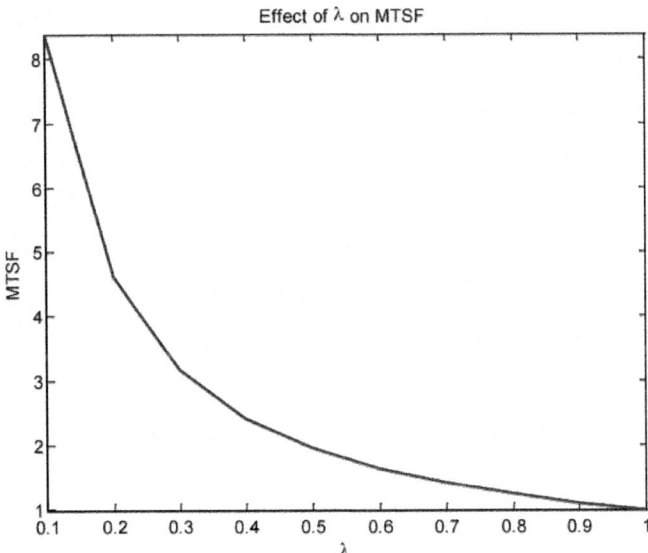

Figure 9: effect of λ on MTSF

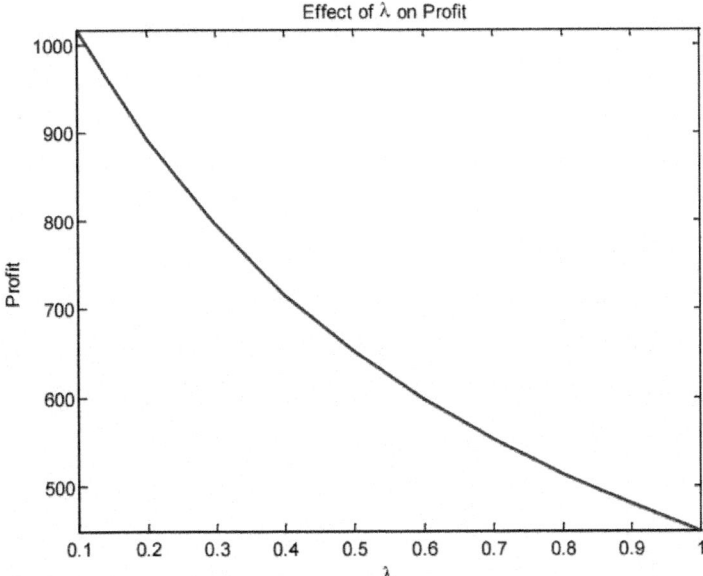

Figure 10: effect of λ on Profit

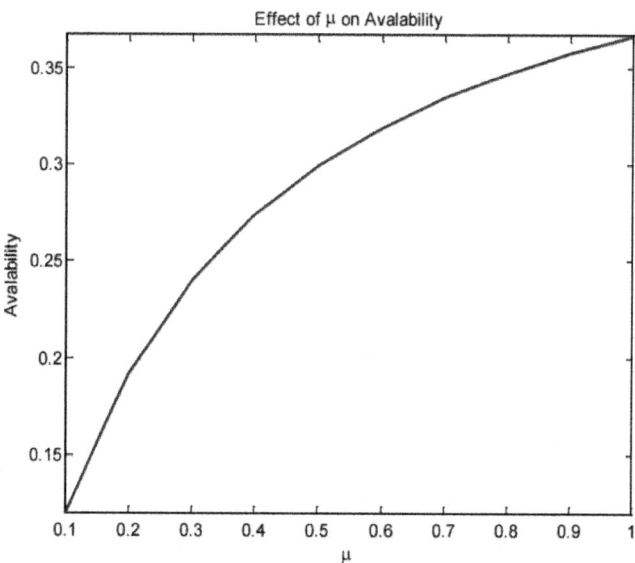

Figure 11: effect of μ on system availability

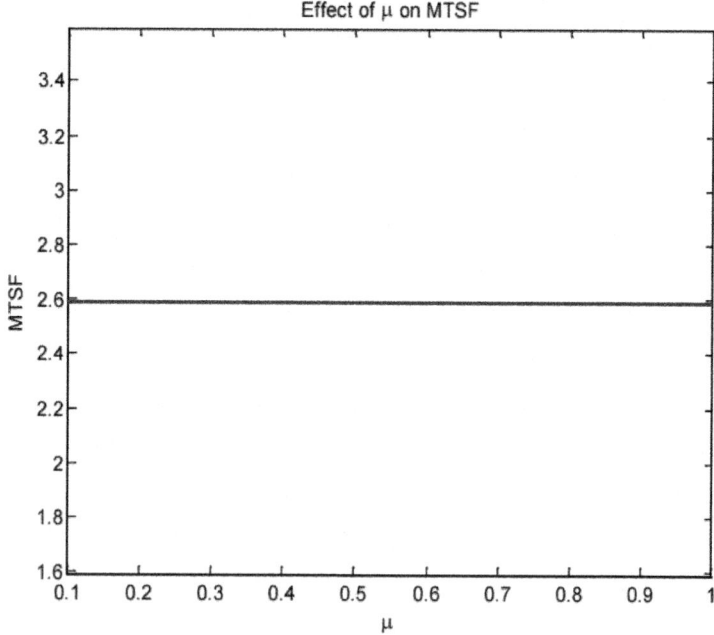

Figure 12: effect of μ on MTSF

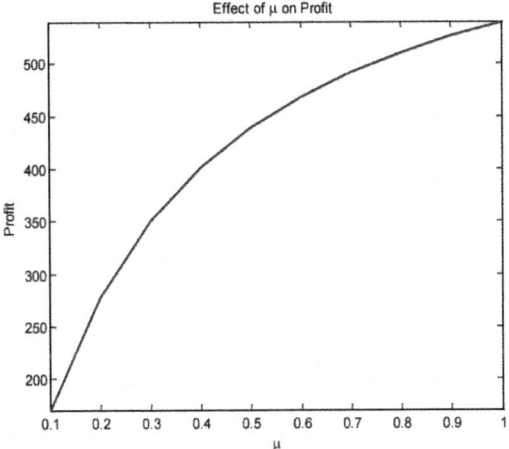

Figure 13: effect of μ on Profit

DISCUSSION

Case I:

$\alpha_2 = 0.5$, $\alpha_3 = 0.02$, $\beta_1 = 0.5$, $\beta_2 = 0.06$, $\beta_3 = 0.01$, $\lambda = 0.6$, $\mu = 0.5$, $C_0 = 1000$, $C_1 = 10$ and vary α_1 for Figure 2 to 4.

Case II:

$\alpha_2 = 0.01$, $\alpha_3 = 0.02$, $\alpha_1 = 0.01$, $\beta_2 = 0.06$, $\beta_3 = 0.9$, $\lambda = 0.06$, $\mu = 0.05$, $C_0 = 1500$, $C_1 = 10$ and vary β_1 for Figure 5 to 7.

Case III:

$\alpha_1 = 0.09$, $\alpha_2 = 0.5$, $\alpha_3 = 0.02$, $\beta_1 = 0.05$, $\beta_2 = 0.06$, $\beta_3 = 0.01$, $\mu = 0.5$, $C_0 = 1500$, $C_1 = 10$ and vary λ for Figure 8 to 10.

Case IV:

$\alpha_1 = 0.5$, $\alpha_2 = 0.5$, $\alpha_3 = 0.02$, $\beta_1 = 0.5$, $\beta_2 = 0.06$, $\beta_3 = 0.01$, $\lambda = 0.6$, $C_0 = 1500$, $C_1 = 10$ and vary μ for Figure 11 to 13.

Figure 2 to 4 provides description of profit function, MTSF and system availability with respect to α_1. From these figures, it is clear that both profit function, MTSF and system availability increase as α1 increases. In Figure 5 to 7, the behavior of system availability, profit function and MTSF are shown with respect to β_1. It is observed that system availability decrease as β_1 increases. In Figure 8 to 10, the behavior of system availability, MTSF and profit function with respect to λ. The results in these figures have shown that system availability, MTSF and profit function decrease as λ increases. Figure 11 to 13 provides description of system availability, MTSF and profit function with respect to μ. System availability and profit in Figure 11 and 13 increase with increase in μ while MTSF in Figure 12 is constant with respect to μ.

CONCLUSIONS

In this paper, we developed the explicit expressions for the mean time to system failure (MTSF), system availability, busy period and profit function for the system and performed graphical study to see the behavior of failure rates and repair rates parameters on system performance. It is observed that from graphical study system performance increase with repair rates and decrease with failure rates.

REFERENCES

1.   Bhardwj, R.K. and Chander, S. (2007). Reliability and cost benefit analysis of 2-out-of-3 redundant system with general distribution of repair and waiting time. DIAS- Technology review- An Int. J. of business and IT. 4(1), 28-35

2.   Chander, S. and Bhardwaj, R.K. (2009). Reliability and economic analysis of 2-out-of-3 redundant system with priority to repair. African J. of Maths and comp. sci, 2(11), 230-236.

3.   Bhardwj,R.K., and S.C. Malik. (2010). MTSF and Cost effectiveness of 2-out-of-3 cold standby system with probability of repair and inspection. Int. J. of Eng. Sci. and Tech. 2(1), 5882-5889

4.   Taneja, G., A. Goyal and D.V. Singh (2011). Reliability and Cost benefit analysis of a system comprising one big unit and two small identical units with priority for operation/repair to big unit. Mathematical Sciences, Vol. 5 No. 3, 235-248

5.   Malik, S.C., Bharwaj, R.K. and Grewal, A.S. (2010). Probabilistic analysis of a system of two non identical parallel units with priority to repair to repair subject to inspection. Journal of reliability and statistical studies, vol. 3(1), pp 1-11

6.   Mahmoud, M.A.W. and Moshref, M.E. (2010). On a two unit cold standby system considering hardware, human error failures and preventive maintenance, Mathematics and Computer modeling, 51(5-6), pp 736-745.

7. Yusuf, I. and Bala, S.I. (2012). Stochastic modeling of a two unit parallel system under two types of failures. International Journal of Latest trends in Mathematics, Vol. 2 (1), pp 44-53

8. Kumar, J. and Kadyan, M.S. (2012). Profit analysis of a system of non identical units with degradation and replacement. International journal of computer application, vol. 40 (3), pp 19-25

9. Sureria, J.K., Malik, S.C. and Anand, J. (2012). Cost benefit analysis of a computer system with priority to software replacement over hardware repair. Applied Mathematical Sciences, vol. 6 (75), pp 3723-3734.

10. El-Said, K.M.,. (2008).Cost analysis of a system with preventive maintenance by using Kolmogorov's forward equations method. Ame. J. of App. Sci. 5(4), 405-410

11. Haggag, M.Y., (2009). Cost analysis of a system involving common cause failures and preventive maintenance.J. Maths. And Stat. 5(4), 305-310

12. El-Said, K.M., and M.S. El-Sherbeny. (2005).Profit analysis of a two unit cold standby system with preventive maintenance and random change in units. J. Maths and Stat.,1 (1) 71-77

13. El-Said, K.M., and M.S. El-Sherbeny. (2005) .Evaluation of reliability and availability characteristics of two different systems by using linear first order differential equations. J. Maths and Stat.,1 (2) 119-123

14. Wang, K., Hseih, C., and Liou, C. H. (2006). Cost benefit analysis of series systems with cold standby components and a repairable service station. Quality technology and quantitative management. Vol. 3(1), pp 77-92.

15. Bhardwaj., R.K., and S.C. Malik (2010). MTSF and cost effectiveness of 2-out-of-3 cold standby system with probability of repair and inspection. International Journal of engineering, science and technology, vol. 2(10), 5882-5889

Chapter 10

ON THE NUMERICAL SOLUTION OF ELLIPTIC AND PARABOLIC OPTIMAL CONTROL PROBLEMS

Francesco Mezzadri, Emanuele Galligani

Department of Engineering "Enzo Ferrari", University of Modena and Reggio Emilia, Strada Vignolese 905/A, I-41125, Modena, Italy

ABSTRACT

This paper regards the solution of some optimal control problems, transcribed as constrained optimization problems, with the main purpose of analysing how the solution changes when the number of discretization points is modified. The problems studied, which include both boundary and distributed elliptic optimal control problems as well as parabolic control problems, have in fact been written in AMPL language and solved by changing the size of the discretization mesh, so to analyse the variance of the minimum of the objective as the number of discretization points increases. Moreover, the problems have been solved using different solvers (such as MINOS, LOQO, IPOPT, SNOPT and KNITRO) and different kinds of discretization so to consider the influence of these parameters on the final results.

INTRODUCTION

This work is concerned with the analysis of the variation of the solution of some optimal control problems as the number of discretization points increases. For this purpose, several optimal control problems have been transcribed as constrained optimization problems which have been then formulated in AMPL language and solved considering different solvers and discretizations. To obtain results as general as possible, both elliptic optimal control problems and parabolic optimal control problems (with Dirichlet, Neumann or mixed boundary conditions) have been studied. In Section 2 the formulation of these problems is therefore presented; in particular, elliptic boundary control problems are first described, distinguishing between the three possible cases of Dirichlet, Neumann and mixed boundary conditions. A general elliptic distributed control problem with mixed boundary conditions is then presented, followed by an example of a parabolic control problem. Eventually, some considerations on the consistency of the discretized solution with the continuous one are discussed, with particular regard to parabolic problems.

Section 3 is instead devoted to the description of the problems actually studied, which have been taken from[1], [2],[3],[4],[5].

In Section 4 the numerical results are presented and discussed. In fact, the complete results obtained with a number of discretization points ranging from 99 to 499 are reported there for every solver and for every considered discretization. The effectiveness of each solver in solving the problems is then sinthetically discussed, together with a list of the errors which more frequently occurred. Eventually, particular effort is given to the analysis of how the solution changes when the number of discretization points increases with the intent of determining if any general trend can be observed.

Finally, Section 5 summarizes the conclusions following from the analysis of the results.

STATEMENT OF THE PROBLEMS

The formulation of elliptic boundary control problems, elliptic distributed control problems and parabolic control problems is here presented; the main references for this section are[6],[7],[1],[2].

Let $\Omega \in \mathbb{R}^2$ be a bounded domain with piecewise smooth boundary Γ; the formulations of the problems are the following.

Elliptic Boundary Control Problems

The formulation of the problems is slightly different depending on the kind of boundary conditions given. In fact, the following three cases can be distinguished:

Elliptic boundary control problem with Dirichlet boundary conditions

$$F(y,u) = \int_{\Omega} f(x, y(x)) dx + \int_{\Gamma} g(x, u(x)) dx \qquad (1)$$

subject to the elliptic state equation

$$-\Delta y(x) + d(x, y(x)) = 0 \qquad for \ x \in \Omega \qquad (2)$$

to the Dirichlet boundary condition

$$y(x) = b(x, u(x)) \qquad\qquad for\ x \in \Gamma \tag{3}$$

and to inequality constraints on control and state

Control: $C(x, u(x)) \le 0 \qquad for\ x \in \Gamma$ (4)

State: $S(x, y(x)) \le 0 \qquad for\ x \in \Omega$ (5)

The functions $f: \Omega \times \mathbb{R} \to \mathbb{R}$, $g: \Gamma \times \mathbb{R} \to \mathbb{R}$, $d: \Omega \times \mathbb{R} \to \mathbb{R}$, $b: \Gamma \times \mathbb{R} \to \mathbb{R}$, $C: \Gamma \times \mathbb{R} \to \mathbb{R}$ and $S: \Omega \times \mathbb{R} \to \mathbb{R}$ are here assumed to be C^2 functions.

2. Elliptic boundary control problem with Neumann boundary conditions

In this case, different boundary conditions are used and some functions are defined on different domain. In fact, the problem consists in determining a boundary control function $u \in L^\infty(\Gamma)$ which minimizes the cost functional

$$F(y, u) = \int_\Omega f(x, y(x)) dx + \int_\Gamma g(x, y(x), u(x)) dx \tag{6}$$

subject to the elliptic state equation (2), to the Neumann boundary condition

$$\partial_v y(x) = b(x, y(x), u(x)) \qquad for\ x \in \Gamma \tag{7}$$

and to inequality constraints on control and state

Control: $C(x, y(x), u(x)) \le 0 \quad for\ x \in \Gamma$ (8)

State: $S(x, y(x)) \le 0 \qquad for\ x \in \bar{\Omega}$ (9)

where $\bar{\Omega} = \Omega \cup \Gamma$ and ∂_v denotes the derivative in the direction of the outward unit normal v of Γ.

The functions $f: \Omega \times \mathbb{R} \to \mathbb{R}$, $g: \Gamma \times \mathbb{R}^2 \to \mathbb{R}$, $d: \Omega \times \mathbb{R} \to \mathbb{R}$, $b: \Gamma \times \mathbb{R}^2 \to \mathbb{R}$, $C: \Gamma \times \mathbb{R}^2 \to \mathbb{R}$ and $S: \bar{\Omega} \times \mathbb{R} \to \mathbb{R}$ are here assumed to be C^2 functions.

3. Elliptic boundary control problem with mixed boundary conditions

In this last case, even the cost function has to be written in a different way: in fact, we have to distinguish between the regions of the boundary where Dirichlet conditions are applied and the ones where we have Neumann conditions instead. The formulation of the problem is therefore the following.

Determine a boundary control function $u \in L^\infty(\Gamma)$ which minimizes the cost functional

$$F(y,u) = \int_{\Omega} f\big(x, y(x)\big) dx + \int_{\Gamma_\alpha} g\big(x, y(x), u(x)\big) dx$$
$$+ \int_{\Gamma_\beta} K\big(x, u(x)\big) dx \tag{10}$$

where $\Gamma_\alpha, \Gamma_\beta \subset \Gamma$ are disjoint sets that consist of smooth and connected components and such that $\Gamma_\alpha, \Gamma_\beta \subset \Gamma$, subject to the elliptic state equation (2), to the mixed Neumann- Dirichlet boundary conditions

$$\partial_v y(x) = b_1\big(x, y(x), u(x)\big) \qquad for\ x \in \Gamma_\alpha \tag{11}$$

$$y(x) = b_2\big(x, u(x)\big) \qquad for\ x \in \Gamma_\beta \tag{12}$$

and to the inequality constraints on control (4) and state (5).

The functions

$f: \Omega \times \mathbb{R} \to \mathbb{R}$, $g: \Gamma_\alpha \times \mathbb{R}^2 \to \mathbb{R}$, $K: \Gamma_\beta \times \mathbb{R} \to \mathbb{R}$, $d: \Omega \times \mathbb{R} \to \mathbb{R}$, $b_1: \Gamma_\alpha \times \mathbb{R}^2 \to \mathbb{R}$, $b_2: \Gamma_\beta \times \mathbb{R} \to \mathbb{R}$,

$C: \Gamma \times \mathbb{R}^2 \to \mathbb{R}$ and $S: \Omega \times \mathbb{R} \to \mathbb{R}$ are here assumed to be C^2 functions.

Elliptic Distributed Control Problem

In this case, the control function u is not on the boundary, but it is distributed in the inner space of the domain. Therefore, supposing that the boundary is partitioned as $\Gamma_\alpha \cup \Gamma_\beta = \Gamma$, where $\Gamma_\alpha, \Gamma_\beta \subset \Gamma$ are disjoint sets that consist of smooth and connected components, the elliptic distributed problem with Neumann and Dirichlet boundary conditions consists in determining a control function $u \in L^\infty(\Omega)$ which minimizes the cost functional

$$F(y,u) = \int_{\Omega} f\big(x, y(x), u(x)\big) dx + \int_{\alpha} g\big(x, y(x)\big) dx \tag{13}$$

subject to the elliptic state equation

$$-\Delta y(x) + d\big(x, y(x), u(x)\big) = 0 \quad for\ x \in \Omega \tag{14}$$

to the Neumann and Dirichlet boundary conditions

$$\partial_v y(x) = b_1\big(x, y(x)\big) \qquad for\ x \in \Gamma_\alpha \tag{15}$$

$$y(x) = b_2(x) \qquad for\ x \in \Gamma_\beta \tag{16}$$

and to inequality constraints on control and state

$$Control: \quad C\big(x, y(x), u(x)\big) \le 0 \quad for\ x \in \Omega \tag{17}$$

$$State: \quad S\big(x, y(x)\big) \le 0 \qquad for\ x \in \overline{\Omega} \tag{18}$$

Where $\bar{\Omega} = \Omega \cup \Gamma_a$.

The functions $f: \Omega \times \mathbb{R}^2 \to \mathbb{R}, \ g: \Gamma_a \times \mathbb{R} \to \mathbb{R}, \ d: \Omega \times \mathbb{R}^2 \to \mathbb{R}, \ b_1: \Gamma_a \times \mathbb{R} \to \mathbb{R}, \ C: \Omega \times \mathbb{R}^2 \to \mathbb{R}$ and $S: \bar{\Omega} \times \mathbb{R} \to \mathbb{R}$ are here assumed to be C^2 functions and $b_2 \in C^2(\Gamma_\beta)$.

Parabolic Control Problem

This type of optimal control problem can be formulated as follows.

Determine a piecewise continuous function $u(x,t)$, $(x,t) \in [0,1] \times [0,1]$ which minimizes the cost functional

$$F(u) = \int_0^1 \int_0^1 \left(y^2(x,t) + \alpha u^2(x,t)\right) dx dt + \int_0^1 y^2(x,1) dx \tag{19}$$

subject to a parabolic equation, such as the diffusion-convection equation

$$y_t = a y_{xx} + b y_x - c y + d + \tilde{\sigma} u \quad (x,t) \in (0,1) \times [0,1] \tag{20}$$

to the initial condition

$$y(x,0) = y_0(x) \qquad\qquad x \in [0,1] \tag{21}$$

to the boundary conditions

$$y(0,t) = g_0(t), \ \ y(1,t) = g_1(t) \quad t \in [0,1] \tag{22}$$

and to inequality constraints on control and state

$$y_{min} \leq y(x,t) \leq y_{max} \quad (x,t) \in [0,1] \times [0,1] \tag{23}$$

$$u_{min} \leq u(x,t) \leq u_{max} \quad (x,t) \in [0,1] \times [0,1] \tag{24}$$

Both the state and the control variables are here functions of space and time. Moreover, a, b and c are known and constant parameters, with $a > 0$ and $c \geq 0$. On the other hand, $\tilde{\sigma}$ can be constant or a given function of the state y. Eventually, d is the source term, which can be a function of y, and $y_0(x)$, $g_0(t)$ and $g_1(t)$ are given functions which satisfy the compatibility conditions

$$y_0(0) = g_0(0) \quad y_0(1) = g_1(0) \tag{25}$$

Consistency

The solution of an optimal control problem often implies its discretization; therefore, the discretization scheme affects the solution itself and it is then important that the optimality conditions of the discretized problem reflect the ones of the continuous problem, as illustrated in[8, Chapt. 4]. It is thus important to analyse the consistency of the optimality conditions to obtain a good discretization of the problem. This can be done by considering the relationship between the Karusk-Kuhn-Tucker conditions, which constitute the discrete necessary conditions, and the continuous necessary conditions, given by the Pontryagin Maximum Principle.

For example, if we consider the parabolic problem defined by the equations (19)-(22), there must be a correspondence between the Karusk-Kuhn-Tucker equations and the co-state equations for the Hamiltonian in the Pontryagin Maximum Principle, thing that often happens when the differential equations are discretized with a difference scheme of low order[9],[10]. Moreover, in[11] the differential equation of this problem is approximated with a finite difference scheme of type FTCS, while the integrals are evaluated using the trapezoidal rule with regard to space integration and using the rectangular method for time integration; this approach results particularly suitable for the stability and the consistency of this problem.

With regard to the studied problems, necessary conditions for the elliptic problems are described in[12] and the verification of second order optimality conditions for the elliptic problems and for the first six parabolic problems is studied in[13]. It is easy to prove that the same considerations can be applied to PC_7 and PC_8 as well.

DESCRIPTION OF THE PROBLEMS

The following problems have been analyzed:

Elliptic Boundary Control Problems (EBC)

1. Problem EBC_1

In this problem, the cost functional consists in the tracking function

$$F(y,u) = \frac{1}{2}\int_{\Omega}(y(x)-y_d(x))^2\,dx$$
$$+\frac{\alpha}{2}\int_{\Gamma}(u(x)-u_d(x))^2\,dx \tag{26}$$

Where $y_d \in C(\bar{\Omega})$ and $u_d \in L^{\infty}(\Gamma)$ are given functions and α is a non negative weight. Moreover, the following constraints and data are given:

on Ω:
$$-\Delta y(x) = 20$$
$$y(x) \leq 3.5$$
$$y_d(x) = 3 + 5x_1(x_1 - 1)x_2(x_2 - 1)$$
on Γ:
$$y(x) = u(x)$$

$$0 \leq u(x) \leq 10$$
$$u_d(x) = 0$$
$$\alpha = 0.01$$

2. Problem EBC_2

This problem is the same as the previous one, with the only exception that α is zero instead of 0.01.

3. Problem EBC_3

The objective is always the tracking function, but the following data have been used:

4. Problem EBC_4

This problem is the same as EBC_3, but with $\alpha = 0$.

5. Problem EBC_5

In this problem (and in the following EBC problems), Neumann boundary conditions have been used. The objective is here the tracking function and the following data are given:

on Ω:
$$-\Delta y(x) = 0$$
$$y(x) \leq 2.071$$
$$y_d(x) = 2 - 2(x_1(x_1 - 1) + x_2(x_2 - 1))$$
on Γ:
$$\partial_v y(x) = u(x) - y(x)^2$$
$$3.7 \leq u(x) \leq 4.5$$
$$u_d(x) = 0$$
$$\alpha = 0.01$$

6. Problem EBC_6

This problem is the same as EBC_5, but some data are different: in fact, we here have $y(x) \leq 2.835$, $6 \leq u(x) \leq 9$ and $\alpha = 0$.

7. Problem EBC_7

Respect to the previous problems, the main difference is that the elliptic state equation is nonlinear; in particular, the following data are given:

on Ω: $-\Delta y(x) - y(x) + y(x)^3 = 0$

$y(x) \le 2.7$

$y_d(x) = 2 - 2(x_1(x_1 - 1) + x_2(x_2 - 1))$

on Γ: $\partial_\nu y(x) = u(x)$

$1.8 \le u(x) \le 2.5$

$u_d(x) = 0$

$\alpha = 0.01$

8. Problem EBC_8

This problem is the same as EBC_7, but with $\alpha = 0$.

Elliptic Distributed Control Problems (EDC)

1. Problem EDC_1

As in the previous problems, also here the objective is the tracking function. The following data are given:

on Ω: $-\Delta y(x) - y(x) + y(x)^3 = u$

$y(x) \le 0.185$

$1.5 \le u(x) \le 4.5$

$y_d(x) = 1 + 2(x_1(x_1 - 1) + x_2(x_2 - 1))$

on Γ: $y(x) = 0$

$u_d(x) = 0$

$\alpha = 0.001$

2. Problem EDC_2

This problem is the same as EDC_1, but with $\alpha = 0$.

3. Problem EDC_3

The problem is the same as EDC_1, but with different data:

4. Problem EDC_4

Even here, the only difference is that different data have been used; in particular, the main difference from EDC_3 is that we have Neumann boundary conditions in place of the Dirichlet ones:

on Ω:
$$- \Delta y(x) - e^{y(x)} = u$$
$$y(x) \le 0.371, -8 \le u(x) \le 9$$
$$y_d(x) = \sin (2\pi x_1)\sin (2\pi x_2)$$

on Γ:
$$\partial_v y(x) + y(x) = 0$$
$$u_d(x) = 0$$
$$\alpha = 0.001$$

5. Problem EDC_5

This problem is the same as EDC_4, but with $\alpha = 0$.

6. Problem EDC_6

In this case, the objective is

$$F(y,u) = \int_\Omega \left(Mu(x)^2 - Ku(x)y(x) \right) dx \tag{27}$$

subject to

1. Elliptic state equation: $-\Delta y(x) = y(x)(a(x) - u(x) - by(x)), \quad x \in \Omega$

2. Omogeneous Neumann boundary conditions: $\partial_v y(x) + \beta(x)y(x) = 0, \quad x \in \Gamma$

3. Constraints on control and state: $1.4 \le u(x) \le 1.6, \quad y(x) \le 6.09, \quad x \in \Omega$

Moreover, the following numerical values are given:

$$a(x) = 7 + 4\sin(2\pi x_1 x_2), \quad b = 1, \quad M = 1, \quad K = 0.8$$
$$\beta(x) = 1 \text{ for } x_1 = 0 \text{ or } x_2 = 0, \text{ else } \beta(x) = 0$$

Parabolic Control Problems (PC)

1. Problem PC_1

The cost functional is here given by

$$F(y,u) = \frac{1}{2}\int_0^l (y(x,T) - y_T(x))^2 \, dx + \frac{\alpha}{2}\int_0^T u(t)^2 dt$$
$$+ \int_0^T \left(a_y(t)y(l,t) + a_u(t)u(t) \right) dt$$

Where $\alpha > 0$ and it is subject to

$$y_t - y_{xx} = 0 \quad in \ (0,l) \times (0,T)$$
$$y(x,0) = a(x) \quad in \ (0,l)$$
$$y_x(0,t) = 0 \quad in \ (0,T)$$
$$y_x(l,t) + \beta y(l,t) = b(t) + u(t) - \varphi(y(l,t)) \quad in \ (0,T)$$
$$\alpha_1 \le u(t) \le \alpha_2 \quad in \ (0,T)$$
$$\alpha_3 \le y(x,t) \le \alpha_4 \quad in \ (0,l) \times (0,T)$$

where:

$$l = \frac{\pi}{4} \qquad T = 1 \qquad \alpha = \frac{\sqrt{2}}{2}\left(e^{2/3} - e^{1/3}\right)$$
$$y_T(x) = (e + e^{-1})\cos x \qquad \alpha_1 = 0 \qquad \alpha_2 = 1$$
$$a(x) = \cos x \quad a_y(t) = -e^{-2t} a_u(t) = \frac{\sqrt{2}}{2} e^{1/3}$$
$$b(t) = \frac{1}{4}e^{-4t} - \min\left(1, \max\left(0, \frac{e^t - e^{1/3}}{e^{2/3} - e^{1/3}}\right)\right)$$
$$\varphi(y) = y|y|^3 \qquad \beta = 1$$

2. Problem PC_2

The problem is the same as PC_1, but with the following numerical data:

$$l = 1 \qquad T = 1.58 \qquad \alpha = 0.001$$
$$y_T(x) = 0.5 \cdot (1 - x^2)\alpha_1 = -1 \qquad \alpha_2 = 1 \qquad \alpha_3 = 0$$
$$\alpha_4 = 1 \qquad a(x) = 0 \qquad a_y(t) = 0 \qquad a_u(t) = 0$$
$$b(t) = 0 \qquad \varphi(y) = 0 \qquad \beta = 1$$

3. Problem PC_3

The only difference from PC_2 is that here we have $\varphi(y) = y^2$ and $\beta = 0$.

4. Problem PC_4

This problem is the same as PC_3, but with $\alpha_3 = 0$ and $\alpha_4 = 0.675$.

5. Problem PC_5

This problem is the same as PC_2, with the following differences: we here use the nonstationary Burgers equation $y_t = v y_{xx} - y y_x$ in $(0,l) \times (0,T)$, with $v = 0.01$, and we have $\alpha_1 = 0.1$ and $\alpha_2 = 0.6$.

6. Problem PC_6

PC_6 is the same as PC_2, but with $\alpha = 0.05$, $\alpha_3 = -1$ and $\alpha_4 = 1$.

7. Problem PC_7

This problem is based on the river aeration problem presented in [4], where more detailed information about the model and the variables are provided. The objective is:

$$F(y,u) = \int_0^{t_f} \int_0^{x_f} Q(x)[y_s(x,t) - y(x,t)]\,dxdt + \lambda \int_0^{t_f} \int_a^b R(x)u^2(x,t)dxdt$$

where the control variable u is the rate of aeration, the state variable y is the concentration of dissolved oxygen, y_s represents the saturation value of dissolved oxygen (here assumed constant and equal to 6 mg/l), while $Q(x)$ and $R(x)$ are weight functions assumed to be equal to 1, just like λ which is a scalar parameter.

The objective is subject to the following constraints:

$$L_t = \frac{1}{A}(AEL_x)_x - vL_x - K_1L + W \ \ in \ (0,x_f) \times (0,t_f)$$

$$y_t = \frac{1}{A}(AEy_x)_x - vy_x - K_2y + U - K_1L + K_2y_s in \ (0,x_f) \times (0,t_f)$$

$$L(x,0) = L_0(x) \qquad in \ (0,x_f)$$

$$y(x,0) = y_0(x) \qquad in \ (0,x_f)$$

$$L(0,t) = L_1(t) \qquad in \ (0,t_f)$$

$$y(0,t) = y_1(t) \qquad in \ (0,t_f)$$

Where L is the biochemical oxygen demand and W is the discharge rate of the biochemical oxygen demand (assumed constant in the analysed problem). Moreover, neglecting the tidal action and assuming constant the cross sectional area, the problem can be simplified by omitting the terms $1/A\,(AEL_x)_x$ and $1/A\,(AEy_x)_x$.

Finally, the following numerical data are given:

$t_f = 5\,days \qquad x_f = 5\,miles \qquad a = 0 \qquad b = x_f$

$K_1 = 0.16\,days^{-1} K_2 = 0.66\,days^{-1} \qquad W = 10$

$v = 1\,mile/day\ L_0(x) = L_1(t) = 0$

$y_0(x) = y_1(t) = y_s = 6\,mg/l$

8. Problem PC_8

PC_8 is based on[5] and regards the problem of firing industrial kilns according to a given reference profile. The objective is:

$$\int_0^T \left[(y(1,t) - p(t))^2 + \alpha u^2(t) \right] dt$$

where $y(x,t)$ is the state variable, $u(t)$ is the control, $T = 0.7$ and $p(t)$, in the numerical example considered, is given by:

$$\begin{cases} 0.7t & if \ 0 \le t \le 0.25 \\ 0.175 & if \ 0.25 \le t \le 0.45 \\ 0.175 - 0.7(t - 0.45) & if \ 0.45 \le t \le 0.7 \end{cases}$$

Moreover, the following constraints are given:

$$Cy_t - (\lambda y_x)_x = q(x,t) \qquad in \ (0,1) \times (0,T)$$

$$\lambda(0,t)y_x(0,t) = y(0,t) - u(t) \qquad in \ (0,T)$$

$$\lambda(1,t)y_x(1,t) = 0 \qquad in \ (0,T)$$

$$y(x,0) = y_0(x) \qquad in \ (0,1)$$

$$y(x,t) \le y_{max} \qquad in \ (0,1) \times (0,T)$$

where:

$$C = C(y) = 4 + y \qquad \lambda = \lambda(y) = 4 - y$$

$$y_0 = 0 \qquad q = 0 \qquad \alpha = 10^{-4}$$

RESULTS AND ANALYSIS

The following results have been obtained by writing the previously presented problems in AMPL language and by solving them with different solvers and number of discretization points. In particular, the problems have been solved through the NEOS servers*http://www.neos-server.org/neos/*[14],[15],[16] by calling the solvers MINOS[17], KNITRO[18], IPOPT[19], LOQO [20] and SNOPT[21], [22].

As regards the discretization, the elliptic problems and the first six parabolic problems have at first been solved with discretizations based on the codes in[23]. Therefore, the elliptic problems have been discretized with the composite rectangle rule to approximate the integrals and with simple discretization techniques (such as the five-point stencil and the Euler method) to approximate the derivatives, while in the first six parabolic problems and in PC_8 the composite trapezoid method, the Crank-Nicolson method and a second-order three-point formula derived from the method of undetermined coefficients have been used. On a second stage of the analysis, other discretizations have then been analysed, as described later in this section.

PC_7, instead, has been discretized by using the FTBS (*forward time backward space*) method for the formulation approximate the integrals; moreover, being the used method explicit, the stability criterion for the method itself has been verified before computing the solution.

In Table 1, then, the results obtained for the problems discretized as described before are presented. The solutions computed by MINOS are not reported since the solver has been able to solve only a few problems, due to the relatively high minimum number of discretization points considered; a short comment on the behaviour of MINOS and of the other solvers is however given after Table 1.

Moreover, it has to be noticed that whenever a second discretization index m was required, it has been set equal to n; for this reason, the only discretization index reported in Tabel 1 is n.

Table 1: Results for the Minimum of the Objective for Different Solvers and Discretization Points

Problem	n	LOQO	IPOPT	SNOPT	KNITRO
EBC_1	99	0.19652520	0.19652519	0.19652520	0.19652521
	199	0.20085512	0.20077162	0.20077162	0.20077164
	299	0.20219539	0.20219539	/	0.20219544
	399	0.20290856	0.20290856	/	0.20339385
	499	0.20333683	0.20333684	/	0.21636265
EBC_2	99	0.09669507	0.09669506	0.09669507	0.09669509
	199	0.10044221	0.10044220	0.10044221	0.10044223
	299	0.10170114	0.10170115	/	0.10170125
	399	0.10233242	0.10233244	/	0.10464184
	499	0.10353798	0.10271179	/	0.10500972
EBC_3	99	0.32100965	0.32100994	0.32100965	0.32105174

	199	0.32812152	0.32812215	0.32812152	0.32826510
	299	0.35194064	0.33050957	/	0.33057592
	399	0.33170556	0.33170361	/	0.33172027
	499	/	0.33242205	/	0.33243834
EBC_4	99	0.25433499	0.24917942	0.24917847	0.24919294
	199	0.26071473	0.25587850	0.25587655	0.25589230
	299	0.25812462	0.25812760	/	0.25818924
	399	0.26190092	0.25925539	/	0.25951613
	499	/	0.25993368	/	0.26143362
EBC_5	99	0.55224624	0.55224693	0.55224875	0.55227779
	199	0.55436880	0.55437238	/	0.55440303
	299	0.55067397	0.55508546	/	0.55519407
	399	/	0.55545525	/	0.55569253
	499	/	0.55569682	/	/
EBC_6	99	0.01507867	0.01507903	3.52315688	0.01508721
	199	0.01560200	0.01560249	/	0.01575581
	299	/	0.01578304	/	0.01603457
	399	/	0.01586932	/	0.03625952
	499	/	0.01593623	/	/
EBC_7	99	0.26416254	0.26416283	3.58730889	0.26417336
	199	0.26728344	/	3.61493172	0.26726350
	299	0.26832619	/	/	0.26831306
	399	0.26884855	/	/	0.26892493
	499	0.26916120	/	/	0.28964582
EBC_8	99	0.16553110	0.16553206	3.52315688	0.16568375
	199	0.16778056	/	/	0.16795923
	299	0.16854248	/	/	0.16898022
	399	/	/	/	0.17200087
	499	0.16915376	/	/	0.16968757
EDC_1	99	0.06216148	0.06216393	0.06263395	0.06227489
	199	0.06442591	0.06445287	0.07892838	0.06477443
	299	0.06519254	0.06530514	0.08003533	0.06592117
	399	0.06557809	0.06570958	/	0.06642852
	499	0.06581013	0.06613078	/	/
EDC_2	99	0.05644742	0.05647171	0.05645288	0.05717146
	199	0.05869679	0.05879243	0.05869960	0.06013342
	299	0.06019334	0.05966575	/	0.06235338
	399	0.05984239	0.06019184	/	0.06275076
	499	0.06007320	0.06058836	/	/
EDC_3	99	0.10988359	0.10988928	0.12499521	0.11288605
	199	0.10988887	0.10990993	0.12499999	0.11417578
	299	0.10988985	0.11003778	0.12499999	0.12266448
	399	0.10989018	0.11002186	0.12499999	0.12370297
	499	0.10989036	0.11006285	0.12499999	/
	99	0.07806397	0.07806781	0.11772134	0.07858685
	199	0.07842594	0.07845272	0.17284174	0.09853755

EDC_4	299	0.07854995	0.07875183	0.18345663	0.10534743
	399	0.07861255	0.07892920	/	0.11122107
	499	0.07865030	0.07899630	/	/
EDC_5	99	0.05447454	0.05449899	0.05468579	0.05701667
	199	0.05642660	0.05488010	0.05919715	0.06979701
	299	/	0.05511081	/	0.07394777
	399	0.05560111	0.05533431	/	0.10859592
	499	0.05611716	0.05557961	/	/
EDC_6	99	-4.2748538	-4.2748368	-4.2742210	-4.2747060
	199	-4.3168810	-4.3168133	/	-4.3154016
	299	-4.3308230	-4.3306766	/	-4.3290439
	399	-4.3377808	-4.3375046	/	/
	499	-4.3419507	-4.3415630	/	/
PC_1	99	/	2.72334553	/	2.72334613
	199	/	2.72156305	/	2.72156405
	299	/	2.72097578	/	2.72097736
	399	/	2.72068361	/	2.72068613
	499	/	2.72050560	/	/
PC_2	99	/	0.00065433	0.00065425	0.00066351
	199	/	0.00065701	/	0.00067061
	299	/	0.00066268	/	/
	399	/	0.00067051	/	/
	499	/	0.00067817	/	/
PC_3	99	0.00052489	0.00051389	0.00051384	0.00053683
	199	/	0.00051383	/	0.00054066
	299	/	0.00051995	/	0.00055214
	399	/	0.00052568	/	0.00054602
	499	/	0.00053191	/	/
PC_4	99	0.00059198	0.00051391	0.00051728	0.00051730
	199	/	0.00051387	/	0.00052878
	299	/	0.00051505	/	0.00055123
	399	/	0.00052849	/	0.00058830
	499	/	0.00053898	/	/
PC_5	99	0.06637097	0.06637136	0.06637099	0.06637123
	199	0.06637202	0.06637103	/	0.06637176
	299	/	0.06637131	/	0.06637189
	399	/	0.06637142	/	0.06637203
	499	/	0.06637142	/	0.06637441

PC_6	99	0.01502941	0.01502942	0.01502941	0.01502943
	199	0.01499875	0.01499876	/	0.01499892
	299	/	0.01499201	/	0.01499212
	399	/	0.01498929	/	0.01500144
	499	0.01498782	0.01498789	/	0.01498872
PC_7	99	62.269023	62.269024	/	62.269025
	199	62.172941	62.172744	/	62.173042
	299	/	62.138290	/	62.139871
	399	/	62.120633	/	/
	499	/	/	/	/
PC_8	70	0.01192334	0.01073357	/	0.01073442
	140	0.01144824	0.01108474	/	0.01108479
	210	0.01131037	0.01116341	/	0.01116343
	280	0.01129613	0.01118859	/	0.01118869
	420	0.01101159	0.01121009	/	/

To enhance the table readability, it has been chosen not to write in the table the kind of error that has been encountered every time the solver was not able to solve the problem, but simply to mark the corresponding box with a slash; the most common errors encountered with each solver are therefore here synthetically described.

MINOS was found not to be fit to solve the vast majority of the problems (even if, actually, it has been possible to solve several problems lowering the number of n below 99, which, however, is the minimum number of discretization points considered in this work). The most common errors are *infeasible problem, infeasible problem (or bad starting guess), too many major iterations, singular basis after several factorization attempts* and *unbounded (or badly scaled) problem.*

The most common errors in LOQO are due to the exceeding of the iteration limit, set to a maximum of 500 iterations, and of the time limit (even though this error occurred more rarely and only for large numbers of discretization points). The reaching of the iteration limit, however, does not mean by itself that the computed solution is far from the real one, since the solver could be near to the convergence to the solution when the limit of 500 iterations was reached; therefore, even when this error occurred, sometimes LOQO computed acceptable solutions, consistent with the ones computed by the other solvers.

In IPOPT the most frequently encountered error is *not enough memory.* Moreover, it has been encountered twice also the error *converged to a locally infeasible point. Problem may be infeasible.* In particular, this happened in EBC_7 and EBC_8 for n=199, where SNOPT converges to the same infeasible point too.

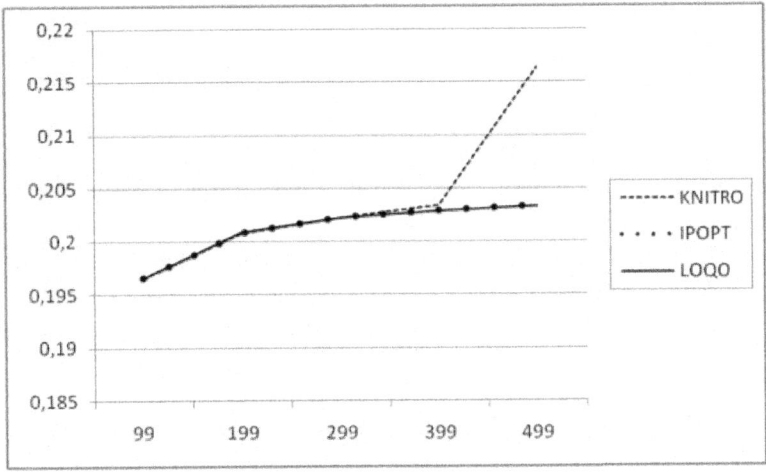

Figure 1: Results for the minimum of the objective for EBC_1

As mentioned in the previous paragraph, SNOPT sometimes computes a solution not in line with the ones computed by the other solvers, as it can be seen in Table 1; moreover, the following errors have often occurred:*not enough real storage, not enough integer storage* and *time limit reached.*

Eventually, the most common error encountered in KNITRO is *not enough memory.*

As regards the variance of the minimum of the objective as the number of discretization points increases, it must be distinguished between elliptic and parabolic problems.

As for the elliptic problems, the value of the objective tends to increase (in absolute value) as the number of discretization points increases, as shown in Table 1. This trend, though, sometimes is not verified for large numbers of discretization points (in fact it is not always possibile to find a definite trend over 399 points). Notwithstanding this, every analysed problem shows an increasing trend at least for $n \leq 299$ for every solver able to properly solve the problem itself. Moreover, even when the increasing trend is not respected in one solver, it is generally so in the others (see Figure 1). Therefore, even if, on one hand, the influence of the solver on the solution seems to be negligible in the vast majority of cases (since the solution often presents only small variations), on the other the analysis of the consistency of the solution between different solvers is of the utmost importance to detect possible out-of-range values. Instead, it is not possible to find any general trend for the parabolic problems: in fact a decreasing trend can be observed in some cases, but it does not happen always, even for a number of discretization points significantly large. Therefore, it appeared more interesting to deepen the study of elliptic problems by verifying if the observed trend is dependent on the type of discretization used; to do so, the elliptic problems have been solved again using different interpolation formulas to approximate the integrals. In particular, while the results reported in Table 1 have been obtained using the composite rectangle method, the ones in Table 2 and in Table 4 have been calculated with a different kind of composite rectangle method, hereafter referred to as *Composite Rectangle Method - B*, evaluating the function on the right bound of each interval rather than on the left one. Eventually, the results in Table 3 and in Table 5 have been obtained using the composite trapezoidal rule.

Table 2: Results Obtained for EDC_1 Discretizing the Integrals with Composite Rectangle Method - B

Problem	*n*	*LOQO*	*IPOPT*	*SNOPT*	*KNITRO*
	99	0.06239886	0.06240544	0.06291330	0.06251088
	199	0.06448084	0.06451681	0.06500796	0.06483747
EDC_1	299	0.06521463	0.06532628	0.06574649	0.06592820
	399	0.06558916	0.06571932	/	0.06642319
	499	0.06583082	0.06613404	/	/

Table 3: Results Obtained for EDC_1 Discretizing the Integrals with the Composite Trapezoidal Rule

Problem	*n*	*LOQO*	*IPOPT*	*SNOPT*	*KNITRO*
	99	0.06230747	0.06231402	0.06279612	0.06242690
	199	0.06446052	0.06449659	0.06497401	0.06481268
EDC_1	299	0.06520681	0.06531911	0.08004900	0.06592788
	399	0.06558546	0.06571649	/	0.06642663
	499	0.06581441	/	/	/

Table 4: Results Obtained for EBC_1 Discretizing the Integrals with Composite Rectangle Method - B

Problem	n	LOQO	IPOPT	SNOPT	KNITRO
EBC_1	99	0.19444468	0.19444468	0.19444468	0.19444513
	199	0.19990545	0.19990545	0.19990545	0.19990557
	299	0.20175738	0.20169022	/	0.20169035
	399	0.20269276	0.20256822	/	0.20673018
	499	0.20314386	0.20308852	/	0.20308883

Table 5: Results Obtained for EBC_1 Discretizing the Integrals with the Composite Trapezoidal Rule

Problem	n	LOQO	IPOPT	SNOPT	KNITRO
EBC_1	99	0.19649294	0.19649294	0.19649294	0.19649429
	199	0.20075226	0.20075227	0.20075226	0.20075234
	299	0.20218429	0.20218429	/	0.20218635
	399	0.20290149	0.20290149	/	0.20340107
	499	0.20336569	/	/	/

Every elliptic problem has been solved with both these discretizations; however, since the solutions are really similar to the ones already reported in Table 1, it has been chosen to present only the results concerning two case problems. The results and the analysis hereafter produced are therefore to be intended as referred to the whole set of elliptic problems and not to EBC_1 and to EDC_1 only.

The tables presented confirm the results previously observed: in fact, though of course there are some differences due to the different discretizations used, the increasing trend is preserved. Therefore it can be concluded that this trend does not depend on the type of discretization used.

It is also worth noticing that the discretization with the trapezoidal rule resulted more stable than the other interpolation formulas: in Table 4, in fact, it can be noticed that for $n = 399$ KNITRO computes a solution which, although it is not so far from the expected value, is larger than and not completely consistent with the solutions computed by LOQO and IPOPT for the same number of discretization points; because of this, the result computed by KNITRO for $n = 499$ is lower than the one for $n = 399$, while in the other solvers the increasing trend is preserved. In Table 5, on the other hand, the solutions show a more regular increasing trend, without any value out of range. This also happens in other problems and, therefore, not only the trapezoidal rule appears to be more stable but the increasing trend is also preserved more often than with the other discretizations. However, it has also to be noticed that the use of the trapezoidal rule also leads to a more complex formulation of the problem, and because of this in Table 5 (and in a lot of other problems discretized this way) it has not been possible to find the solution for $n = 499$ with the solvers IPOPT and KNITRO.

As regards the errors occurred, the same considerations made for Table 1 apply; the only thing that can be noticed is that the error *Time limit* is more frequent in solving the problem with the integrals discretized with the trapezoidal rule due to the greater computational complexity of the method itself. For the same reason, also the error *not enough memory* occurs more frequently and for lower numbers of discretization points.

CONCLUSIONS

The results presented and analysed in the previous section allow to come to the following conclusions:

1. As for elliptic problems, it is possible to identify an increasing trend of the absolute value of solution as the number of discretization points increases. This trend can be observed in every elliptic problem studied for a number of discretization points n lower than or equal to 299. Instead, for $n = 399$ and for $n = 499$ the trend may or may not be respected for some solvers.

2. The use of different solvers or of different interpolation formulas to approximate the integrals of the objective changes the value of the solution, yet it does not influence the trend described in the previous point.

3. It is not possible to identify any general trend as regards the parabolic problems: in fact, in PC_1 and in PC_7 the solution decreases as n increases, while in other problems an increasing trend is observed and in others it is not possible to identify any trend at all.

4. Regarding the parabolic problems, the trend can be dependent on the solver used: for example in PC_4 the solution computed with KNITRO increases, while the one computed with IPOPT at first decreases and then increases. In general, the solutions computed with KNITRO tend to increase as n increases for $n \leq 299$ with the sole exception of PC_1, while IPOPT has less regular trends and the other solvers are often not able to properly solve the problem.

REFERENCES

1. H. Maurer, and H. D. Mittelmann, "Optimization techniques for solving elliptic control problems with control state and constraint: part 1. Boundary control", Computational Optimization and Applications, vol. 16, pp 29-55, 2000.

2. H. Maurer, and H. D. Mittelmann, "Optimization techniques for solving elliptic control problems with control state and constraint: part 2. Distributed control", Computational Optimization and Applications, vol. 18, pp 141-160, 2001.

3. H. D. Mittelmann, "Sufficient Optimality for Discretized Parabolic and Elliptic Control Problems", International Series of Numerical Mathematics, vol. 138, pp. 184–196, 2001.

4. W. Hullett, "Optimal Estuary Aeration: an Application of Distributed Parameter Control Theory", Applied Mathematics & Optimization, vol. 1, pp. 20-63, 1974.

5. F. Leibfritz, E. W. Sachs, "Numerical Solution of Parabolic State Constrained Control Problems Using SQP- and Interior-Point-Methods", Large Scale Optimization: State of the Art, (W.W. Hager et al. eds.), pp. 245–258, 1994.

6. E. Galligani, "Iterative Solution of Elliptic Control Problems with Control and State Constraints", Atti del Seminario Matematico e Fisico dell'Università di Modena e Reggio Emilia,vol. 53, 2, pp. 365-408. 2005.

7. S. Bonettini and V. Ruggiero. (2006)A Collection of Optimal Control Problems with Control and State Constraints.[Online] Available:http://dm.unife.it/~bonettini/ip_pcg/controllo_1.pdf.

8. J. T. Betts, "Practical Methods for Optimal Control Using Nonlinear Programming", SIAM, Philadelphia, 2001.

9. E. Galligani, "On Solving Optimal Control Problems for Distributed Parameter Systems Using an Inexact Newton Method", Recent Advances in Nonlinear Optimization and Equilibrium Problems: a Tribute to Marco D'Apuzzo, Quaderni di Matematica, vol. 27, (V. De Simone, D. di Serafino, G. Toraldo eds.), Dipartimento di Matematica, Seconda Università degli Studi di Napoli, pp. 227-252, 2012.

10. C. T. Kelley, "Iterative Methods for Optimization", SIAM, Philadelphia, 1999.

11. A.C. Hindmarsh, P.M. Gresho, and D.F. Griffith, "The stability of explicit Euler time-integration for certain finite difference approximations of the multi-dimensional advection-diffusion equation", International Journal

for Numerical Methods in Fluids, vol. 4, pp. 853-897, 1984.

12. H. D. Mittelmann, H. Maurer, "Solving Elliptic Control Problems with Interior Point Methods: Control and State Constraints", Journal of Computational and Applied Mathematics, 120, pp. 175-195, 2000.

13. H. D. Mittelmann, "Verification of Second-Order Sufficient Optimality Conditions for Semilinear Elliptic and Parabolic Control Problems", Computational Optimization and Applications, 20, pp. 93–110, 2001.

14. M. Mesnier, J. Morè, and J. Czyzyk, "The NEOS Server", IEEE Journal on Computational Science and Engineering, pp. 68-75, 1998.

15. J. Morè and W. Gropp, "Optimization Environments and the NEOS Server", Approximation Theory and Optimization, pp. 167-182, 1997.

16. E. D. Dolan, "The NEOS Server 4.0 Administrative Guide", Mathematics and Computer Science Division, Argonne National Laboratory, Tech. Mem. 250, 2001.

17. B. A. Murtagh and M. A. Saunders, "MINOS 5.5 user's guide", Department of Operations Research, Stanford University, Tech. Rep. 83-20R, 1998.

18. J. Nocedal, R. A. Waltz, and R. H. Byrd, "KNITRO: An Integrated Package for Nonlinear Optimization", Large Scale Nonlinear Optimization, pp. 35-59, 2006.

19. O. Schenk, H. D. Simon, S. Toledo, and U. Naumann, "Short Tutorial: Getting Started with IPOPT in 90 Minutes", Dagstuhl Seminar Proceedings 09061, Combinatorial Scientific Computing, 2009.

20. R. J. Vanderbei, "LOQO User's Manual – Version 4.05", Department of Operations Research and Financial Engineering, Princeton University, Princeton, 2006.

21. W. Murray, M. A. Saunders, and P. E. Gill, "SNOPT: An SQP Algorithm for Large-Scale Constrained Optimization", SIAM Review, vol. 47, no. 1, pp. 99-131, 1997.

22. W. Murray, M. A. Saunders, and P. E. Gill, "User's Guide for SNOPT: a Fortran Package for Large-Scale Nonlinear Programming", Tech. Rep. SOL96-0, Department of Mathematics, University of California, San Diego, 1996.

23. H. D. Mittelmann. (2002) A Collection of Test Problems in PDE-Constrained Optimization.[Online]. Available:http://plato.asu.edu/pdecon.html

CITATION

CHAPTER 1

Edeki, S. , Okagbue, H. , Opanuga, A. and Adeosun, S. (2014) A Semi-Analytical Method for Solutions of a Certain Class of Second Order Ordinary Differential Equations. Applied Mathematics, 5, 2034-2041. doi: 10.4236/am.2014.513196.

CHAPTER 2

V. Ananthaswamy and L. Rajendran, "Analytical Solutions of Some Two-Point Non-Linear Elliptic Boundary Value Problems," Applied Mathematics, Vol. 3 No. 9, 2012, pp. 1044-1058. doi: 10.4236/am.2012.39154.

CHAPTER 3

P. Savenko, "Computational Methods in the Theory of Synthesis of Radio and Acoustic Radiating Systems," Applied Mathematics, Vol. 4 No. 3, 2013, pp. 523-549. doi: 10.4236/am.2013.43078.

CHAPTER 4

Olabode B. T., Momoh A. L. Continuous Hybrid Multistep Methods with Legendre Basis Function for Direct Treatment of Second Order Stiff ODEs doi:10.5923/j.ajcam.20160602.03

CHAPTER 5

Yusuf A., Aiyesimi Y. M., Jiya M., Okedayo G. T., Analysis of Couette Flow of a Nanofluid in an Inclined Channel with Soret and Dufour Effects, American Journal of Computational and Applied Mathematics , Vol. 6 No. 2, 2016, pp. 57-64. doi: 10.5923/j.ajcam.20160602.05.

CHAPTER 6

Drakos Stefanos, Constitutive Relations of Stress and Strain in Stochastic Finite Element Method, American Journal of Computational and Applied Mathematics , Vol. 5 No. 6, 2015, pp. 164-173. doi: 10.5923/j.ajcam.20150506.02.

CHAPTER 7

Paul T. R. Wang, Generalized Gaussian Elimination (GGE) Solving System of Linear Inequalities or Equalities (LIS-II), American Journal of Computational and Applied Mathematics , Vol. 4 No. 6, 2014, pp. 202-217. doi: 10.5923/j.ajcam.20140406.04.

CHAPTER 8

Emanuele Galligani, "The Arithmetic Mean Solver in Lagged Diffusivity Method for Nonlinear Diffusion Equations", American Journal of Computational and Applied Mathematics, Vol. 2 No. 6, 2012, pp. 241-248. doi: 10.5923/j.ajcam.20120206.01.

CHAPTER 9

Ibrahim Yusuf, Nafiu Hussaini, Stochastic Modeling of Repairable Redundant System Comprising One Big Unit and Three Small Dissimilar Units, American Journal of Computational and Applied Mathematics, Vol. 2 No. 4, 2012, pp. 174-188. doi: 10.5923/j.ajcam.20120204.06.

CHAPTER 10

Francesco Mezzadri, Emanuele Galligani, On the Numerical Solution of Elliptic and Parabolic Optimal Control Problems, American Journal of Computational and Applied Mathematics , Vol. 3 No. 5, 2013, pp. 259-268. doi: 10.5923/j.ajcam.20130305.02.

INDEX